Mathematical
Theory of
Electrophoresis

Mathematical Theory of Electrophoresis

V. G. Babskii
M. Yu. Zhukov
V. I. Yudovich

Institute of Molecular Biology and Genetics
Academy of Sciences of the Ukrainian SSR
Kiev, USSR

Translated from Russian by

Cathy Flick

CONSULTANTS BUREAU
NEW YORK AND LONDON

Library of Congress Cataloging in Publication Data

Babskiı, V. G. (Vitaliĭ Genrikhovich)
  [Matematicheskaía teoriía elektroforeza. English]
  Mathematical theory of electrophoresis / V. G. Babskiĭ, M. Yu. Zhukov, and
V. I. Yudovich; translated from Russian by Cathy Flick.
    p.   cm.
  Translation of: Matematicheskaía teoriía elektroforeza.
  Bibliography: p.
  ISBN-13:978-1-4612-8225-9   e-ISBN-13:978-1-4613-0879-9
  DOI:10.1007/978-1-4613-0879-9

  1. Electrophoresis—Mathematical models. I. Zhukov, M. IŪ. (Mikhail
IÙr'evich) II. IÙdovich, V. I. (Viktor Iosifovich) III. Title.
QP519.9.E434B3313   1988   88-25282
574.87′072—dc19   CIP

This translation is published under an agreement with the Copyright
Agency of the USSR (VAAP)

© 1989 Consultants Bureau, New York
Softcover reprint of the hardcover 1st edition 1989

A Division of Plenum Publishing Corporation
233 Spring Street, New York, N.Y. 10013

PREFACE

The development of contemporary molecular biology with its growing tendency toward in-depth study of the mechanisms of biological processes, structure, function, and identification of biopolymers requires application of accurate physicochemical methods. Electrophoresis occupies a key position among such methods. A wide range of phenomena fall under the designation of electrophoresis in the literature at the present time. One common characteristic of all such phenomena is transport by an electric field of a substance whose particles take on a net charge as a result of interaction with the solution. The most important mechanisms for charge generation are dissociation of the substance into ions in solution and formation of electrical double layers with uncompensated charges on particles of dispersed medium in the liquid.

As applied to the problem of separation, purification, and analysis of cells, cell organelles, and biopolymers, there is a broad classification of electrophoretic methods primarily according to the methodological characteristics of the process, the types of supporting media, etc. An extensive literature describes the use of these methods for the investigation of different systems. A number of papers are theoretical in nature. Thus, the microscopic theory has been developed rather completely [13] by considering electrophoresis within the framework of electrokinetic phenomena based on the concept of the electrical double layer.

Having begun to work with several problems in electrophoresis, we observed that most questions of practical importance which are of interest to experimenters are macroscopic in nature. Examples are the formation of zones with elevated amounts of individual components, the degree of separation, phenomena connected with diffusion and heat evolution, and the distribution of the electric field and charges. Such questions may be answered only by studying the physicomechanical properties of the

v

medium in which electrophoresis is carried out, the interactions of individual components of the medium with each other and with the electric field, chemical reactions, and heat and mass transport. All these processes are closely interconnected and cannot be studied in isolation.

We have become convinced that electrophoretic theory should be based on equations describing the hydrodynamics, electrodynamics, and thermodynamics of multicomponent, chemically reactive mixtures, i.e., on the most general equations of the physics of continuous media. In this case, we no longer need phenomenological verification of a specific electrophoresis model, and, in particular, we eliminate the possibility of omitting various cross-coupled phenomena. To our knowledge, there are no such general equations in the literature (see, for example, [11, 19, 30, 32, 34, 43]). The derivation of these equations as presented in Chapter I is based, as usual, on application of the methods of nonequilibrium thermodynamics [3, 10, 11, 14, 38, 54]. We hope that these equations will be useful far beyond the range of phenomena connected with electrophoresis. The extreme complexity of this system practically eliminates its direct use due to its generality. However, such a system may serve as a basis for constructing mathematical models describing different specific situations. Such simplifications decrease the dimensionality of the system and are based on comparative analysis of the importance of different effects. Secondary effects are discarded and the important ones taken into account using asymptotic methods.

In the literature, the process usually begins with special models, from which we go to the more general ones. In our view, a method based on going from the most general model to a specific one has a number of advantages. There is no need to go over the entire route each time the balance equations for mass, energy, etc., are derived. Reliable control over the assumptions made in the derivation of a specific model and the conditions for its applicability are achieved. The greater flexibility of the approach is apparent in the possibilities for improving the model by more accurate accounting of different effects.

An unavoidable limitation of the physics of continuous media and, in particular, the thermodynamics of irreversible processes is the presence in the basic equations of transport coefficients (viscosity, thermal conductivity, diffusion, mobility, etc.) which must be determined from experiment. The difficulties are aggravated if we need to know the dependence of these coefficients on the thermodynamic parameters (temperature, pressure, concentration, electric and magnetic field, etc.). In principle, these dependencies should be described by a different (microscopic) theory, perhaps from statistical physics, but at the present time these dependencies are known only in isolated cases. The experimental difficulties arising in this case are obvious: it is easy to determine the parameter, it is more complicated to obtain the experimental curve, and it is practically impossible to determine how the transport coefficients depend on two or more thermodynamic parameters.

The general approach developed in this monograph allows us to determine the relevant relations for multicomponent mixtures from the general equations using an asymptotic method which separates fast and slow variables. Thus, we can determine the concentration dependence of the mobility and the electrical conductivity, as specified by the chemistry of a given multicomponent mixture. Finally, the form of the material relations is determined only for pseudocomponents, chemical complexes formed as a result of chemical reactions in solution. The determination of transport coefficients for the individual components included in the complex still remains unsolved. It is also important that with the deductive approach it is easy to discover diverse connections and parallels between different specific models. Thus, we perceive a profound internal commonality in the phenomena and models of isotachophoresis, zone electrophoresis, and isoelectric focusing.

In constructing specific models, it was found that the extant classification of types of electrophoresis is often inadequate. Expansion of the methodological nomenclature often follows explicit commercial goals. A meaningful classification scheme should obviously be derived within the framework of the theory of electromigration phenomena, and be based primarily on the composition and properties of the buffer systems and mixtures to be separated. For example, when the concentrations of the substances to be separated are comparable with the buffer concentration, zone electrophoresis transforms into isotachophoresis. Also, if the isoelectric point of a specific substance falls within the proper pH range, zone electrophoresis and isoelectric focusing may be carried out simultaneously in the chamber. Nevertheless, for convenience in comparison with known results, we have decided to adhere to generally accepted terminology.

The current theory makes it possible to describe the time evolution of electrophoretic processes, including such phenomena as the rearrangement of zones and the establishment of a steady state in isotachophoresis, diffusion dispersion and the creation of meniscus-forming zones in zone electrophoresis, and the evolution of an artificial pH gradient and zones close to the isoelectric point in isoelectric focusing. In addition, known theoretical relationships, which usually pertain to the steady state, acquire a more profound physical interpretation and prove to be special cases of the universal patterns for these phenomena. Thus, the Kohlrausch relation for a concentration jump in isotachophoresis proves to be no different than the Hugoniot condition across a shock wave.

The concept of an "infinite-component" mixture developed in this work is important. It allows us to describe phenomena, such as the creation of a natural pH gradient in a mixture with a very large number of ampholine-type carrier ampholytes, which are at first glance inaccessible to theory. These mixtures are not described by discrete sets of concentrations, but instead by distribution functions in a space of continuous latent parameters.

Work with the models presented in this book has disclosed that applications of the theory are not exhausted by phenomena known from experiment. Within the framework of this theory, we have obtained a number of new qualitative results, e.g., the phenomena of electrolytic memory and electrolytic mixing observed in the study of isotachophoresis. The former is connected with the appearance of an absolute Riemann invariant, a function of concentration conserved over time for each point in the solution. The second phenomenon is described from the mathematical point of view as a progressive rarefaction wave.

In constructing the theory, we have tried to consider all experimental data available to us. The subsequent fate of the theory depends on the existence of an inverse relationship with experiment. We need a thorough test of its qualitative and quantitative predictions, measurement of the parameters introduced, and estimates of the relative roles of various effects.

In conclusion, we mention some peculiarities of this monograph connected with the fact that this book may arbitrarily be classified as physicochemical biology, and the methods described in it are the methods of the physics of continuous media. Therefore, our physicist readers may lack information from chemistry and biology, while our biologist readers will hardly recognize all the mathematical fine points. Nevertheless, we hope that the attraction of this frontier area, the unusual and unexpected nature of the physical effects, the novelty of the mathematical formulations, and the prospects for biologically interesting applications will reward readers for their efforts.

FOREWORD
TO THE ENGLISH EDITION

Electrophoresis of low-molecular-weight compounds and biopolymers is a dynamic and rapidly developing field both from the standpoint of instrumentation and techniques and from the standpoint of the theory of the process and its mathematical model. Since the publication of the Russian edition of this book, progress has been made in several directions, and new results have been obtained by us and other researchers. We will point out some of these developments in this foreword.

The method for calculating titration curves demonstrated in Chapter III for amino acids with several ionogenic groups has been extended to peptides and proteins with a known primary structure. By specifying the dissociation constants of the ionogenic groups in a given protein, we can determine the approximate pI value and then refine this value by varying the values of the constants and taking into account the secondary and tertiary structure of the protein – where, of course, the inverse problem also arises: determination of the dissociation constants from the titration curve [1].

The non-steady-state model of isotachophoresis presented in Chapter IV has been significantly developed. Explicit formulas that describe the motion of zones have been obtained for weak and strong electrolytes, thus eliminating the somewhat cumbersome reasoning used in constructing the tables and graphs [2–4], and algorithms have been developed for computer calculations. A model for isotachophoresis has been constructed that is free from the restrictions on the composition of the mixture given in Section $\cdot$ 1 of Chapter IV. For reactions of the type $H^+B \rightleftarrows B + H^+$, $HA_i \rightleftarrows A_i^- + H^+$ ($i = 1, \ldots, n$), where H^+B is a base and HA_i are the acids to be separated, algebraic algorithms written for the computer have been constructed which allow one to calculate the motion of the zones, the position of their boundaries at any instant of time, the time for complete separation of the mixture, the concentrations of the substances, and the pH values in the zones. The mixture components HA_i ($i = 1, ..., n$) may be strong or weak acids, or amphoteric substances displaying acidic properties (the pH of the zone in this case should differ considerably from the pI of the substance). The indicated reaction scheme is most typical for experiments when a strong acid (usually HCl) is chosen as the leader, and the mixture consists of amino acids.

A mathematical model for isotachophoresis at constant potential has been constructed [5]. This isotachophoresis method was recommended in [6] for determining the electrophoretic mobility of substances; however, the theory of the process given in [6] describes only a special case that is not realizable experimentally.

Coulophoretic titration has been investigated. This is a process for determining the electrolyte concentration when the zones move in opposing directions [7]. Analysis of the corresponding mathematical model has made it possible to describe [4] the electrosynthesis process: the development in the electrophoretic chamber of moving and fixed regions of electrolytes that did not exist in the solution at the initial instant of time.

Models have also been constructed that take into account diffusion processes in isotachophoresis (a calculation of the width of the boundary in the steady-state case is given in [8]).

Certain advances have been made in describing the motion of individual zones in zone electrophoresis, extending further the results of Chapter V. The effect of the buffer on the mobility of an amphoteric substance present in low concentration has been clarified [9]. It is shown that the ratio of their conductivities plays an important role in the description of the interaction between the buffer and the amphoteric substance. An amphoteric substance found in some local volume together with the buffer displaces buffer ions from this volume. Here, for the case where the conductivity of the amphoteric substance is less than the conductivity of the buffer, a decrease in conductivity of the given volume occurs; consequently, an increase in the electric field occurs within this volume according to the law $E = E_0(1 + \alpha c)$, where E_0 is the electric field in a region free of the amphoteric substance, $c \ll 1$ is the concentration of the amphoteric substance, and α is the weighting coefficient (in this case, $\alpha > 0$; when the conductivity of the amphoteric substance is greater than the conductivity of the buffer, the opposite effect occurs, and $\alpha < 0$). This means that the dependence $v = v_0(1 + \alpha c)$ is satisfied for the electrophoretic transport rate. For the case $\alpha > 0$ (usually for high-molecular-weight compounds), non-linear effects lead to broadening of the trailing edge of the zone: development of a "tail" of the substance behind the moving zone. Conversely, when $\alpha < 0$, the leading edge of the zone is broadened. Calculations show that electromigrational broadening of the zones in most cases occurs significantly earlier than their diffusional broadening, especially when the concentration of the substance in the zone is high [9].

Progress has also been made in constructing models for creating pH gradients in finite-component mixtures, as demonstrated in Chapter VI for borate–polyol systems. It has been shown (see [10]) that in order to carry out accurate calculations we need to detail the reaction scheme (1.1)-(1.4) in Section 1 of Chapter VI, taking into account the effect of the ionic strength and other factors on the dissociation constants. Such additional information is especially important for the traditionally used tris-borate buffer. Furthermore, the theory has been refined for borate–polyol systems by taking into account diffusion processes affecting the quality of the pH gradients [11].

We should also note one promising route for creating a pH gradient: formation of a pH gradient moving at a low constant velocity, using isotachophoresis, based in this case on the rather convincing argument that it is better to have a controlled, constant motion of the pH gradient than to have an uncontrolled cathode or anode drift, or a "plateau" for the pH gradient created by the carrier ampholytes. For example, for a set of weak acids HA_i ($i = 1, ..., n$) and one weak base H^+B used as the counterion, it is not difficult to obtain relations specifying the required pH gradient (see [5]).

With regard to the theory of the creation of pH gradients in infinite-component systems and isoelectric focusing in such pH gradients (Chapter VII), we emphasize again that the theory is based on the fine difference between the pH values at which the mobility (pI_i) and charge (pI_e) of the amphoteric substance go to zero. The amphoteric substance in solution is represented by an ionic complex, the anions and cations of which in the general case have different mobilities. The number, or more precisely the concentration, of the anions and cations in the complex depends on the pH of the medium. When $pH = pI_e$, this does not at all mean that the mobility of the complex in an electric field is equal to zero. The point is that the electric field does not act on the complex as a whole, but rather on its cations and anions individually; as a result, the complex may have nonzero mobility even when $pH = pI_e$. When $pH = pI_i$, the concentration of anions and cations is such that the mobility of the complex is equal to zero. The difference between pI_e and pI_i is significant for low-molecular-weight compounds, in particular for low-molecular-weight carrier ampholytes, and the difference in practice is insignificant for high-molecular-weight compounds such as proteins since, in such proteins, the mobilities of the cations and anions are, as a rule, practically equal.

It was not our intention to write a review of the mathematical theory of electrophoresis, but rather to present the results of our own investigations. However, mention must be made of the productive group of M. Bier in the USA (see, for example [12] and the papers cited therein). Conceptually, these papers are quite similar to our book; however, whereas we accentuate analytical (in particular, asymptotic) methods, the work of Bier et al. [12] is oriented mainly toward numerical modeling as applied to simple model systems (see, for example, [8]). It is evident that these two approaches complement one another and promote a deeper understanding of the electrophoretic processes, especially in those cases where theory is compared with experiment.

In conclusion, we note that the theory that we have presented serves as a basis for constructing models of the development of thermal and concentration convection in electrophoresis. This is especially important in connection with the fact that convection is specifically a major artifact of electrophoresis that reduces its efficiency and resolution. We refer to the work of Zhukov and Korol' [9], where a theoretical and experimental study has been carried out on concentration convection in zone electrophoresis for a model substance in a stabilizing sucrose density gradient.

REFERENCES

1. V. Kašická, J. Vacík, and Z. Prusík, "Determination of dissociation constants of weak electrolytes by capillary isotachophoresis," *J. Chromatogr.*, **320**, 33–43 (1985).
2. M. Yu. Zhukov and V. I. Yudovich, "Mathematical model of isotachophoresis," *Dokl. Akad. Nauk SSSR*, **267**, No. 2, 334–338 (1982).
3. M. Yu. Zhukov, "Nonsteady-state model of isotachophoresis," *Zh. Vychisl. Mat. Mat. Fiz.*, No. 4, 549–565 (1984).
4. M. Yu. Zhukov, "Technique for calculating motion of zones and time for complete separation of a mixture in isotachophoresis," in: *Molecular Biology* [in Russian], Naukova Dumka, Kiev (1984), No. 36, pp. 28–34.
5. M. Yu. Zhukov and L. E. Korol', "Use of isotachophoresis at constant potential for determining mobility," *Biopolim. Kletka*, **2**, No. 5, 257–261 (1986).
6. M. Carchon and E. Eggermont, "Isotachophoretic determination of relative apparent mobility as an estimate of electrophoretic mobility of ions," *Electrophoresis*, No. 3, 263–274 (1982).
7. V. O. Oshurkova and I. A. Ivanova, "Coulophoretic titration," *Dokl. Akad. Nauk SSSR*, **227**, No. 6, 1371–1378 (1976).

8.   W. Thormann and R. A. Mosher, "Theoretical and computer-aided analysis of steady-state moving boundaries in electrophoresis: An analytical solution for the estimation of boundary widths between weak electrolytes," *Trans. Soc. Comput. Simul.*, **1**, No. 1, 83–96 (1984).

9.   M. Yu. Zhukov and L. E. Korol', "Convective, diffusional, and electromigrational zone broadening in free liquid zone electrophoresis," in: *Space Biology and Biotechnology* [in Russian], Naukova Dumka, Kiev (1986), pp. 54–64.

10. B. M. Michov, "Calculation of 'tris-borate' ion mobility," *Electrophoresis*, **5**, 171 (1984).

11. G. Yu. Azhitskii, M. Yu. Zhukov, L. E. Korol', G. V. Troitskii, and V. G. Babskii, "Evolution of a pH gradient in the boric acid–glycerine system," in: *Molecular Biology* [in Russian], Naukova Dumka, Kiev (1984), No. 36, pp. 51–56.

12. M. Bier, O. A. Palusinski, R. A. Mosher, and D. A. Saville, "Electrophoresis: Mathematical modeling and computer simulation," *Science*, **219**, No. 4590, 1281–1287 (1983).

CONTENTS

xv

Chapter I

BASIC EQUATIONS

An electric field may induce unusual transport processes in a continuous medium if the medium contains free charges or is capable of generating free charges under the influence of the field. A great many such media are known, but the nature of the free charge carriers and physical mechanisms of formation are quite diverse. In metals, the electrical charge is transported by free electrons; in semiconductors, the free charge carriers are electrons, "holes," and so-called electron–hole droplets [39]. The characteristic carriers in a plasma are electrons and ions. In liquid electrolytes, charge is transported by ions from dissociating substances.

An interesting range of phenomena in multiphase, multicomponent media in an electric field of importance to practitioners is connected with the generation of an electrokinetic potential at interfaces. Accordingly, these are called electrokinetic phenomena. The study of such phenomena was begun by the Moscow chemist F. F. Reuss (1809), who observed the motion of colloidal particles in solution along the direction of the applied electric field. This phenomenon was called electrophoresis. The elementary theory of electrophoresis given by M. Smoluchowski is based on the idea that an electrical double layer, two or three molecules thick, exists at the interface. For definiteness, consider a solid–liquid interface. The electrical double layer may be likened to a capacitor (Fig. 1), one of whose plates coincides with the interface and is rigidly connected with the solid phase (the liquid particles adhere to the solid); the outside plate (relative to the solid) consists of particles which may slip parallel to the interface if an electric field acts in this direction. If we neglect the curvature of the interface and consider the capacitor to be flat, we may make use of a relation familiar from electrostatics which connects the potential ζ between capacitor plates with the surface charge density σ on the plate:

1

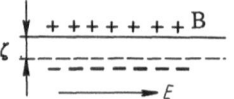

Fig. 1. Electrical double layer.

$$\sigma = \zeta \varepsilon \varepsilon_0,$$

where ε_0 is the dielectric constant of the vacuum, and $\varepsilon \varepsilon_0$ is the dielectric constant of the liquid (so that ε is their ratio). The potential denoted as ζ is the electrokinetic potential (also called the Helmholtz potential, or the ζ potential).

In the following, we assume that the velocity profile of the liquid along the boundary is linear. This allows us to use the expression $\eta v / \delta$ for the tangential viscous stress (the force per unit area), where η is the dynamic viscosity coefficient for the liquid and v is the velocity outside the double layer.

For slow flows inertial effects may be ignored and the velocity is determined from the condition that the driving force and the resisting force are equal. Here, by equating the electric force σE to the viscous drag $\eta v / \delta$ and using the expression given for σ, we obtain the formula

$$v = \gamma E, \quad \gamma = \zeta \delta \varepsilon \varepsilon_0 / \eta.$$

The quantity γ is called the mobility of the particle; it is the velocity per unit field strength. Taking into account the curvature of the interfacial surface within the framework of this theory leads to the appearance of the dimensionless coefficient f ($0 < f \leq 1$) in the formula for the mobility. The formula

$$\gamma = f \zeta \varepsilon \varepsilon_0 \delta / \eta$$

is in better agreement with experiment.

In addition to electrophoresis, the category of electrokinetic phenomena includes the Dorn effect, which is the opposite of electrophoresis. It involves generation of a potential difference in the direction of relative motion of the phases induced by mechanical forces. This phenomenon arises in cuvets, test tubes, and rotors of preparative and analytical ultracentrifuges during sedimentation separation or analysis of macromolecules. Here, the so-called primary charge effect is displayed [34, 69], since the sedimentation of charged macromolecules engenders an electric field which slows down the macromolecular ion and speeds up the counterions. The secondary charge effect [34, 69] is due to unequal mobilities and sedimentation coefficients for the ions of a simple electrolyte in mixed polyelectrolyte–electrolyte solutions. This effect must be taken into account during equilibrium configuration in density gradients with salts of cesium and other heavy metals [98].

Another pair of complementary electrokinetic phenomena are electroosmosis (motion of liquid through capillaries or porous diaphragms under the influence of an internal electric field) and streaming potential (the generation of a potential difference between ends of a capillary through which liquid flows).

The elementary theory provides only a relationship between the quantities γ and ζ. Determination of the ζ-potential requires deeper consideration. A number of papers (see [13]) are devoted to solution of this important problem, which is based on the study of the nonlinear Poisson equation for the electric potential and involves significant mathematical difficulties.

The theory of electrokinetic phenomena, as applied to a solitary particle, may be called a microscopic theory. However, to resolve a number of important questions in electrophoresis we need a macroscopic theory describing the motion of the dispersed phase as a whole. For example, in most cases of practical interest the microscopic theory is inadequate. It does not allow us, for example, to determine the velocity and the shape of the zones in zone electrophoresis and isoelectric focusing, or the degree of separation, or the concentration distribution in the electrophoretic column or the zones. Of course, the velocity of the zone, for example, may be determined by calculating the mobility of its component particles when we know the driving electric field. And the electric field intensity may be determined from the potential difference specified at the ends of the column, if the field is assumed to be homogeneous. But the fact that zones are generated suggests strong inhomogeneities in the field distribution. Consequently, its determination becomes possible only on the basis of sufficiently complete allowance for electrochemical reactions in the column.

In principle, two approaches are possible in constructing the macroscopic theory. The first approach is based on the methods of statistical mechanics. The collection of particles is considered as a gas in which individual particles play the role of molecules. We introduce analogs of average density, temperature, and pressure, and consider their fluctuations. This approach to constructing the theory for a multicomponent continuous medium has been outlined in [1, 18]. One important difficulty in such an approach is connected with correcting for complicated interactions between particles which, in contrast to the ideas of classical statistical mechanics, may be neither potential nor instantaneous in character. We arrive at a definite result by this route only if we admit certain hypotheses relative to the character of the interactions. Choosing correct hypotheses when we have an acute deficiency of experimental data is not an easy matter.

The second approach consists of successive application of the principles of nonequilibrium thermodynamics to the set of problems under consideration. This chapter is devoted to developing this approach. The theory is based on the idea originating from Fick (1855), wherein a multicomponent medium is considered to be a collection of interacting media and fields all moving in the same space. Each of the media is characterized by its own (partial) density, pressure, velocity, etc.

Construction of the theory begins with applying to each component the basic phenomenological laws of physics: conservation of mass, charge, energy, momentum, and angular momentum. Since the systems cannot be considered closed, these laws take the form of the corresponding balance equations. Finally, there are too few balance laws to obtain a closed system of equations; they must be supplemented by defining relations characterizing the physicochemical properties of the medium as a whole. Nonequilibrium thermodynamics gives us a uniform approach to deriving the balance equations and investigating the defining relations. Numerous papers [3, 10, 11, 19, 28, 30, 38, 42–44, 54, 59, 82, etc.] have been devoted to the development of nonequilibrium thermodynamics and the application of its methods to the construction of a theory for multicomponent media.

The interconnection between the phenomenological theories and (still unconstructed) statistical theories of electrophoresis laid out below may be clarified by analogy with fields of physics such as hydrodynamics and thermodynamics. Both approaches should have a broad, common region of applicability and, in this region, they should lead to the same results; the microscopic theory should give additional expressions for the transport coefficients (viscosity, thermal conductivity, diffusion, etc.) in terms of the microscopic quantities. Calculation of the transport coefficients (for example viscosity or thermal conductivity) according to the microscopic theory is a complicated problem, which may be considered unsolved at the present time, even for classical systems. The macroscopic theory assumes that these coefficients are known and proposes recipes for their experimental determination. From what has been said, it follows that the microscopic and macroscopic theories should be developed in parallel.

An advantage of the macroscopic theory of special importance in the study of electrophoresis of biopolymers is the possibility of considering processes in homogeneous and heterogeneous media consistently. For example, electrophoresis in aqueous solutions of proteins should be preferably considered as the migration of ions under the influence of an electric field, instead of the motion of dispersed particles. However, discussion of the similarity and differences in the laws of motion for macromolecules and small dispersed particles remains within the framework of the microscopic theory. The phenomenology of these processes is the same and, we repeat, the type of microscopic interaction only affects the magnitude of the transport coefficients (and they are determined experimentally), and not the form of the macroscopic relationships.

Constructing a complete macroscopic theory for electrophoresis requires simultaneous consideration of a large set of physical and chemical processes and interactions. First of all, of course, we must consider the transport of electric charges under the influence of the electric field. Consequently, we must introduce into the theory the electromagnetic properties of the medium (including polarization) and also the mechanisms for formation of electrical charge carriers, especially chemical reactions connected with dissociation. Also, we need to take into account processes of thermal conduction and diffusion, including cross-diffusion (for example thermal diffusion). Diffusion not only determines the width of the zones in zone electrophoresis and isoelectric focusing, but also exerts an important effect on the electric field distribution and consequently on the motion of the zones in a number of cases. Convection plays an important role in electrophoresis. This role is most often negative, although there are examples of the constructive use of convection in some electrophoretic methods [77].

Thus, we conclude that in order to develop a theory of electrophoresis we need to use more general equations for the physics of continuous media: the equations of hydrodynamics, electrodynamics, and thermodynamics for multicomponent, chemically reactive mixtures [15]. To our knowledge, such general equations have never been derived in the literature, although numerous special cases of them are known (equations for convection, magnetohydrodynamics, dynamics of electrolytes, chemical kinetics, etc.). We should note that these equations take into account almost all conceivable physical effects possible in multicomponent, chemically reactive mixtures. However, the vast number of different components and the diversity of the chemical reactions occurring between components make the equations in the theory very unwieldy, and often practically inaccessible. Moreover, it is obvious that the vast amount of information contained in these equations and their solutions is too detailed for practical use.

We clearly perceive the need for applying the commonly held idea in physics that we should reduce the description so that it includes only information which is in fact important for understanding the phenomena, their practical application, and their control. At the same time, it is *a priori* rather difficult to indicate the conditions under which specific effects may be discarded. Nevertheless, even when deriving a complete system of equations for a multicomponent mixture (Section 14), we use a number of simplifying hypotheses. For example, in the equations of motion we neglect inertial forces arising during motion of the individual components relative to the mixture and the momentum transport during motion of mass sources relative to the mixture. Furthermore, we assume that the magnetic field is weak and slowly varying; we do not take into account the magnetization of the mixture, or nonlinear effects arising upon polarization, or the contribution of electrostriction to the mass fluxes of the components. In the expressions for the mass and heat fluxes, we neglect various cross-effects; in the equation for determining the temperature, we neglect heat sources connected with the work done by the viscous stresses and with the temperature dependence of the partial pressures and dielectric constants.

However, the major simplifications are made in constructing mathematical models for specific electrophoretic phenomena from the complete system of equations. In Chapters III-VII, we formulate the physical conditions under which a specific model is valid. As a rule, approximations such as the weak diffusion hypothesis, the separation and specific allowance for fast and slow reactions, the assumption that the concentration of one or several components of the mixture is small, are used. One important simplification is connected with introducing the idea of an "infinite-component" medium (one which has an infinite number of components) [17]. When there are very many components, it is no longer expedient to investigate the behavior of each component in detail. Instead, attention is given only to the behavior of groups of components with similar parameters. This approach is demonstrated in the last section of this chapter, and then it is used to describe isoelectric focusing.

Such a transition from general equations to models of electrophoresis has allowed us to establish fundamental connections between different types of electrophoresis, in particular between isotachophoresis, zone electrophoresis, and isoelectric focusing in infinite-component systems (see Chapters III-VI). It is precisely the *a priori* physical assumptions made in constructing isolated models for electrophoresis which kept us from seeing these connections considerably earlier. Finally, in the overwhelming majority of papers, different types of electrophoresis are considered as essentially new physical phenomena, which are not interconnected.

1. Mass Balance

Let a multicomponent mixture fill the region D of the three-dimensional space R^3. It is convenient to designate the number of components as $N + 1$. We will characterize each component of the mixture by the density $\rho_k(x, t)$ and the velocity $v_k(x, t)$, where k is the component index ($k = 1, 2, \ldots, N + 1$, and t denotes time. We introduce the density of the mixture $\rho(x, t)$ and the weighted-mean velocity of the mixture, which characterizes the mixture as a whole

$$\rho = \sum_{k=0}^{N} \rho_k, \quad \mathbf{v} = \frac{1}{\rho} \sum_{k=0}^{N} \rho_k \mathbf{v}_k. \tag{1.1}$$

The integral equation for the mass balance of the kth component is represented as

$$\frac{d}{dt} \int_R \rho_k dv + \int_{\partial R} \rho_k (\mathbf{v}_k \cdot \mathbf{n}) \, dA = \int_R \sigma_k dv, \quad k = 0, \ldots, N, \tag{1.2}$$

where R is an arbitrary fixed volume of the region D; $\mathbf{n}$ is the outward normal to the surface ∂R, the boundary to the region R; σ_k is the mass source density for the kth component arising, for example, as the result of chemical reactions.

From (1.2) we derive (applying the Gauss theorem to the surface integral) the differential equation for the mass balance of the kth component:

$$\frac{\partial \rho_k}{\partial t} + \operatorname{div} (\rho_k \mathbf{v}_k) = \sigma_k, \quad k = 0, \ldots, N. \tag{1.3}$$

Summing (1.3) over all components, with allowance for (1.1) we obtain the mass balance equation for the mixture:

$$\frac{\partial \rho}{\partial t} + \operatorname{div} (\rho \mathbf{v}) = \sum_{k=0}^{N} \sigma_k. \tag{1.4}$$

Introducing the material derivatives for the mixture and the kth component,

$$\frac{d}{dt} (\ ) = \frac{\partial}{\partial t} (\ ) + \mathbf{v} \cdot \nabla (\ ), \quad \frac{d_k (\ )}{dt} = \frac{\partial (\ )}{\partial t} + \mathbf{v}_k \cdot \nabla (\ ), \tag{1.5}$$

we rewrite relationships (1.3) and (1.4) in the form

$$\frac{d\rho}{dt} + \rho \operatorname{div} \mathbf{v} = \sum_{k=0}^{N} \sigma_k, \quad \frac{d_k \rho_k}{dt} + \rho_k \operatorname{div} \mathbf{v}_k = \sigma_k. \tag{1.6}$$

Define the mass flux density of the mixture $\mathbf{i}^0$, the mass flux density of the kth component $\mathbf{i}_k^0$, and the diffusion mass flux density of the kth component $\mathbf{i}_k$, setting

$$\mathbf{i}^0 = \rho\mathbf{v}, \quad \mathbf{i}_k^0 = \rho_k\mathbf{v}_k, \quad \mathbf{i}_k = \rho_k(\mathbf{v}_k - \mathbf{v}) \equiv \rho_k\mathbf{w}_k \quad (k = 0, \ldots, N). \quad (1.7)$$

From (1.1) it follows that the mass flux of the mixture consists of the sum of the mass fluxes for the individual components, and there is no overall diffusion mass flux, i.e.,

$$\mathbf{i}^0 = \sum_{k=0}^{N} \mathbf{i}_k^0, \quad \sum_{k=0}^{N} \mathbf{i}_k \equiv \sum_{k=0}^{N} \rho_k\mathbf{w}_k = 0. \quad (1.8)$$

We assume that the law of conservation of mass is satisfied for the mixture:

$$\sum_{k=0}^{N} \sigma_k = 0 \quad (1.9)$$

or, alternatively, that the continuity equation for the mass flux is valid:

$$\frac{\partial \rho}{\partial t} + \operatorname{div}(\rho\mathbf{v}) = 0; \quad \left(\frac{d\rho}{dt} + \rho \operatorname{div}\mathbf{v}\right) = 0. \quad (1.10)$$

Introducing the mass concentrations of the components,

$$c_k = \frac{\rho_k}{\rho} \quad (k = 0, \ldots, N), \quad \sum_{k=0}^{N} c_k = 1, \quad (1.11)$$

we rewrite the mass balance equation for the kth component in (1.3) in the form

$$\rho\frac{dc_k}{dt} + \operatorname{div}\mathbf{i}_k = \sigma_k \quad (k = 0, \ldots, N). \quad (1.12)$$

2. Momentum Balance

We assume the integral momentum balance for the kth component in the form

$$\frac{d}{dt}\int_R \rho_k\mathbf{v}_k dv + \int_{\partial R} \rho_k\mathbf{v}_k(\mathbf{v}_k \cdot \mathbf{n})\, dA = \int_{\partial R} \mathbf{T}_k \cdot \mathbf{n} dA + \int_R \{\sigma_k\mathbf{J}_k + \rho_k\mathbf{F}_k$$
$$+ \rho_k\mathbf{F}_k^e + \rho_k\boldsymbol{\pi}_k\}\, dv. \quad (2.1)$$

Here, $\mathbf{F}_k$ is the specific external force acting on the kth component; $\mathbf{F}_k^e$ is the specific electromagnetic force acting on the kth component; π_k is the specific internal force acting on the component as viewed from the other components and arising, for example, as a result of intermolecular interactions; $\sigma_k \mathbf{J}_k$ is the momentum source density arising as the result of motion of the mass source of the kth component (usually it is assumed that the quantity $\mathbf{J}_k$ coincides with the velocity of the kth component, $\mathbf{J}_k = \mathbf{v}_k$; see the review in [59]); $\mathbf{T}_k$ is the stress tensor for the kth component, in which electromagnetic stresses may be included.

From (2.1) the differential momentum balance equation for the kth component follows as

$$\frac{\partial}{\partial t}(\rho_k \mathbf{v}_k) + \operatorname{div}(\rho_k \mathbf{v}_k \mathbf{v}_k - \mathbf{T}_k) = \sigma_k \mathbf{J}_k + \rho_k(\mathbf{F}_k + \mathbf{F}_k^e + \pi_k). \quad (2.2)$$

It may also be written in the form

$$\rho_k \frac{d_k \mathbf{v}_k}{dt} = \operatorname{div} \mathbf{T}_k + \sigma_k(\mathbf{J}_k - \mathbf{v}_k) + \rho_k(\mathbf{F}_k + \mathbf{F}_k^e + \pi_k) \quad (k = 0, 1, \ldots, N). \quad (2.3)$$

Summing over all components, we obtain the differential momentum balance equation for the mixture:

$$\rho \frac{d\mathbf{v}}{dt} = \operatorname{div} \sum_{k=0}^{N}(\mathbf{T}_k - \rho_k \mathbf{w}_k \mathbf{w}_k) + \sum_{k=0}^{N} \rho_k \mathbf{F}_k + \sum_{k=0}^{N} \rho_k \mathbf{F}_k^e + \sum_{k=0}^{N}(\sigma_k \mathbf{J}_k + \rho_k \pi_k). \quad (2.4)$$

We require that the momentum source, connected with the mass sources for the components, be balanced by the momenta of the internal forces [14, 59].

$$\sum_{k=0}^{N}(\sigma_k \mathbf{J}_k + \rho_k \pi_k) = 0. \quad (2.5)$$

Let us determine the stress tensor for the mixture $\mathbf{T}$, the specific external force $\mathbf{F}$ acting on the mixture, and the specific electromagnetic force $\mathbf{F}^e$ acting on the mixture, by setting (see [11, 14])

$$\mathbf{T} = \sum_{k=0}^{N}(\mathbf{T}_k - \rho_k \mathbf{w}_k \mathbf{w}_k), \quad (2.6)$$

$$\rho \mathbf{F} = \sum_{k=0}^{N} \rho_k \mathbf{F}_k, \quad (2.7)$$

$$\rho \mathbf{F}^e = \sum_{k=0}^{N} \rho_k \mathbf{F}_k^e. \quad (2.8)$$

Then the momentum balance equation for the mixture takes on the form

$$\rho \frac{dv}{dt} = \text{div } T + \rho F + \rho F^e. \tag{2.9}$$

Finally, the relationship obtained represents the well-known equation of motion with respect to the stresses (see, for example, [11, 30]), which may be written directly for the mixture as a whole. The possibility of deriving this relationship from the equations of motion for the individual components indicates that the assumptions are not contradictory.

3. Angular Momentum Balance

In most problems in the physics of continuous media, the stress tensor may be considered to be symmetric. However, this property of a continuous medium is connected with the absence of any internal moments within the medium, and such a property may not apply, for example, to media with magnetic properties. For such media, called "moment-containing media" [42], we must consider separately the balance equation for angular momentum; in a general situation it does not follow from the momentum balance equation.

We will use the integral equation for the angular momentum balance for the kth component of the mixture relative to an arbitrary fixed vector x_0 in the form

$$\frac{d}{dt} \int\limits_R (x - x_0) \wedge \rho_k v_k dv + \int\limits_{\partial R} (x - x_0) \wedge \rho_k v_k (v_k \cdot n)\, dA = \int\limits_R (x - x_0)$$

$$\wedge (\sigma_k J_k + \rho_k F_k + \rho_k F^e + \rho_k \pi_k)\, dv + \int\limits_{\partial R} (x - x_0) \wedge T_k \cdot n dA + \int\limits_R (\varepsilon : \lambda_k) dv,$$

$$(\varepsilon : \lambda_k)^\alpha \equiv \varepsilon_{\alpha\beta\gamma} \lambda_k^{\beta\gamma}, \tag{3.1}$$

where λ_k is the antisymmetric tensor of the internal moments of the kth component; $\varepsilon_{\alpha\beta\gamma}$ is a third-rank tensor which is totally skew-symmetric relative to the indices $\alpha\beta\gamma$.

Using (2.9), we obtain the differential equation for angular momentum balance for the kth component in the form

$$\varepsilon : (T_k^a - \lambda_k) = 0, \tag{3.2}$$

where T_k^a is the antisymmetric part of the stress tensor for the kth component. Summing over all components of the mixture and using (2.6), we derive

$$T^a \equiv \sum_{k=0}^{N} T_k^a = \sum_{k=0}^{N} \lambda_k. \tag{3.3}$$

The latter relationship is the condition which should be satisfied by the stress tensor for the mixture.

4. Equations of the Electromagnetic Field in Matter and Determination of the Ponderomotive Forces

We will describe the electromagnetic field acting on the mixture by the electric field intensity vector $\mathbf{E}$ and the axial magnetic field intensity vector $\mathbf{H}$. As parameters characterizing the electromagnetic properties of the kth component, we will take the specific free charge e_k, the specific dipole moment $\mathbf{p}_k$, and the specific magnetic moment $\mathbf{m}_k$.

We will write Maxwell's equations for the mixture in the form (see, for example, [9, 31, 32, 42, 43, 82])

$$\operatorname{rot} \mathbf{H} = \varepsilon_* \frac{\partial \mathbf{E}}{\partial t} + \sum_{k=0}^{N} \mathbf{j}_k^e + \sum_{k=0}^{N} \mathbf{j}_k^P, \qquad (4.1)$$

$$\operatorname{rot} \mathbf{E} = -\mu_* \frac{\partial \mathbf{H}}{\partial t} - \sum_{k=0}^{N} \mathbf{j}_k^M, \qquad (4.2)$$

$$\mu_* \operatorname{div} \mathbf{H} = \sum_{k=0}^{N} \rho_k^M, \qquad (4.3)$$

$$\varepsilon_* \operatorname{div} \mathbf{E} = \sum_{k=0}^{N} \rho_k^e + \sum_{k=0}^{N} \rho_k^P. \qquad (4.4)$$

Here ε_*, μ_* are the dielectric constant and the magnetic permeability of free space; ρ_k^e, ρ_k^P, ρ_k^M are, respectively, the free charge density, polarization charge density, and magnetization charge density; $\mathbf{j}_k^e$, $\mathbf{j}_k^P$, $\mathbf{j}_k^M$ are the electric current density for the free charges, polarization current density, and the magnetization current density. In this form Maxwell's equations embody the following hypothesis. In the absence of matter, i.e., when all charges and currents are equal to zero, the electromagnetic field can be described using only the two field parameters $\mathbf{E}$ and $\mathbf{H}$. The interaction of the electromagnetic field with matter (for example, polarization and magnetization) may be described using charges and currents; the effect of these charges and currents is equivalent to the effect to which they would lead in free space.

Using Maxwell's equations in form (4.1)-(4.4) is very convenient for describing multicomponent mixtures, since this allows us to graphically represent the contributions of the individual components to the electromagnetic effects arising in the mixture. Finally, the charge and current densities entering into (4.1)-(4.4) must be connected with the quantities e_k, p_k, m_k which we use to describe the electromagnetic properties of the mixture components.

We define the free charge density for the kth component as

$$\rho_k^e = \rho_k e_k. \tag{4.5}$$

Represent the electric current density for free charges for the kth component in the form

$$j_k^e = \rho_k^e v_k. \tag{4.6}$$

Note that in some cases it is impossible or inadvisable to divide the mixture into components such that the electric current density for free charges is connected only with the macroscopic motion of the components. Here we must introduce the conduction current $j_k^{e'}$, adding to it the right-hand side of (4.6). This case is not considered further (see [11]). Let us introduce the polarization vector $\mathbf{P}_k$ and the magnetization vector $\mathbf{M}_k$ (the volume dipole and magnetic moments):

$$\mathbf{P}_k = \rho_k p_k, \quad \mathbf{M}_k = \rho_k m_k. \tag{4.7}$$

Next define the polarization and magnetization charge density for the kth component, having required that the electric and magnetic moments of these charges for the entire volume D occupied by the mixture coincide with the dipole and magnetic moments of the mixture due to the kth component. Furthermore, the total polarization and magnetic charges of the region D for the kth component are zero. Thus,

$$\int_D \rho_k^P \mathbf{x} dv = \int_D \mathbf{P}_k dv, \quad \int_D \rho_k^P dv = 0, \int_D \rho_k^M \mathbf{x} dv = \int_D \mathbf{M}_k dv, \quad \int_D \rho_k^M dv = 0. \tag{4.8}$$

Let us assume that there is no polarization or magnetization outside the region D, and that on the boundary ∂D, the surface charges and currents are equal to zero ($\mathbf{P}_k \cdot \mathbf{n} \equiv \mathbf{M}_k \cdot \mathbf{n} \equiv 0, x \in \partial D$). Then, to satisfy (4.8), it is sufficient to require

$$\rho_k^P = - \operatorname{div} \mathbf{P}_k, \quad \rho_k^M = - \operatorname{div} \mathbf{M}_k. \tag{4.9}$$

We represent the polarization and magnetization current density for the kth component in the form

$$j_k^P = \rho_k^P v_k + j_k^{P'}, \quad j_k^M = \rho_k^M v_k + j_k^{M'}, \tag{4.10}$$

where $\rho_k{}^P v_k$, $\rho_k{}^M v_k$ are the convective current densities, respectively, for polarization and magnetization; $j_k{}^{P'}$, $j_k{}^{M'}$ are the internal current densities, respectively, for polarization and magnetization.

Next we define the internal current densities for polarization and magnetization of the kth component in such a way that the total polarization and magnetization currents of the kth component passing through an arbitrary liquid surface S_k, the change in which is connected only with motion of the kth component, coincide with the change in the flux of the polarization vector P_k and the magnetization vector M_k through the area S_k:

$$\int_{S_k} j_k^{P'} \cdot n dA = \frac{d_k}{dt} \int_{S_k} P_k \cdot n dA, \quad \int_{S_k} j_k^{M'} \cdot n dA = \frac{d_k}{dt} \int_{S_k} M_k \cdot n dA. \tag{4.11}$$

By virtue of the arbitrary nature of S_k, we derive from (4.11), using (4.9),

$$j_k^{P'} = \frac{\partial P_k}{\partial t} + \text{rot}\,(P_k \wedge v_k) - \rho_k^P v_k, \quad j_k^{M'} = \frac{\partial M}{\partial t} + \text{rot}\,(M_k \wedge v_k) - \rho_k^M v_k. \tag{4.12}$$

From (4.9), (4.10), and (4.12) the continuity equations for the polarization and magnetization current density for the kth component follow as

$$\frac{\partial \rho_k^P}{\partial t} + \text{div}\, j_k^P = 0, \quad \frac{\partial \rho_k^M}{\partial t} + \text{div}\, j_k^M = 0. \tag{4.13}$$

We obtain the continuity equations for the free charge current densities from (4.1) and (4.4), taking into account (4.13):

$$\frac{\partial}{\partial t} \sum_{k=0}^{N} \rho_k^e + \text{div} \sum_{k=0}^{N} j_k^e = 0. \tag{4.14}$$

Finally, we write Maxwell's equations (4.1)-(4.4) for the mixture in the form

$$\text{rot}\left(H - \sum_{k=0}^{N} P_k \wedge v_k\right) = \frac{\partial}{\partial t}\left(\varepsilon_* E + \sum_{k=0}^{N} P_k\right) + \sum_{k=0}^{N} (\rho_k e_k v_k), \tag{4.15}$$

$$\text{rot}\left(E + \sum_{k=0}^{N} M_k \wedge v_k\right) = -\frac{\partial}{\partial t}\left(\mu_* H + \sum_{k=0}^{N} M_k\right), \tag{4.16}$$

$$\text{div}\left(\mu_* H + \sum_{k=0}^{N} M_k\right) = 0, \tag{4.17}$$

$$\text{div}\left(\varepsilon_* E + \sum_{k=0}^{N} P_k\right) = \sum_{k=0}^{N} \rho_k e_k. \tag{4.18}$$

Let us introduce the Lorentz force acting on the free charge, the polarization charge, and the magnetization charge of the kth component as

$$\rho_k F_k = \rho_k^e E + \mu_* (j_k^e \wedge H) + \rho_k^P E + \mu_* (j_k^P \wedge H) + \rho_k^M H - \varepsilon_* (j_k^M \wedge E). \quad (4.19)$$

For the following development, it is convenient to introduce the effective electric and magnetic field intensities for the kth component, i.e., the field intensities which are measured in a coordinate system which is fixed relative to the kth component:

$$E_k = E - \mu_* (H \wedge v_k), \quad H_k = H + \varepsilon_* (E \wedge v_k). \quad (4.20)$$

Then the expression for the Lorentz forces is written in the form

$$\rho_k F_k = \rho_k^e E_k + \rho_k^P E_k + \rho_k^M H_k + \mu_* (j_k^{P'} \wedge H) - \varepsilon_* (j_k^{P'} \wedge E). \quad (4.21)$$

In the general case the force on the mixture is not reduced to a Lorentz force alone (see, for example, [5, 9, 32]). To see this, define the energy density of the electromagnetic field $\rho \varepsilon_e$, not including polarization and magnetization energy, by the expression

$$\rho \varepsilon_e = \frac{1}{2} (\varepsilon_* E^2 + \mu_* H^2). \quad (4.22)$$

Write the balance equation for the electromagnetic energy in the form

$$\frac{\partial}{\partial t} \rho \varepsilon_e + \text{div } S = - \sum_{k=0}^{N} \rho_k F_k^e \cdot v_k - \sum_{k=0}^{N} \sigma_k^e, \quad (4.23)$$

where S is the electromagnetic energy flux density; $\rho_k F_k^e$ is the density of electromagnetic forces acting on the mixture; σ_k^e is the electromagnetic energy source density.

The balance equation (4.23) does not include the polarization and magnetization energy, so the forces F_k^e do not completely describe the force effect of the electromagnetic field on the mixture. The polarization and magnetization energies of the kth component will be included in the internal energy of the kth component. In such a description, the forces arising upon polarization and magnetization will enter into the partial pressures of the components of the mixture.

We will use expression (4.23) for known $\rho \varepsilon_e$, S, and σ_k^e as defining the electromagnetic forces acting on the components of the mixture.

Differentiating (4.22), using (4.1)-(4.4), (4.10), (4.19), and (4.20), we obtain

$$\frac{\partial}{\partial t}\,\rho\varepsilon_e = -\,\mathrm{div}\,(\mathbf{E}\wedge\mathbf{H}) - \sum_{k=0}^{N}\rho_k F_k\cdot\mathbf{v}_k - \sum_{k=0}^{N}(\mathbf{j}_k^{\mathbf{P'}}\cdot\mathbf{E}_k + \mathbf{j}_k^{\mathbf{M'}}\cdot\mathbf{H}_k). \qquad (4.24)$$

Transforming this expression while taking into account (4.7), (4.9), and (4.12), we derive

$$\frac{\partial}{\partial t}\,\rho\varepsilon_e = -\,\mathrm{div}\,(\mathbf{E}\wedge\mathbf{H}) - \sum_{k=0}^{N}\rho_k F_k\cdot\mathbf{v}_k - \sum_{k=0}^{N}j_k^{e'}\cdot\mathbf{E}_k$$

$$-\sum_{k=0}^{N}\left\{\left(\rho_k\,\frac{d_k\mathbf{p}_k}{dt} + \mathbf{p}_k\sigma_k\right)\cdot\mathbf{E}_k - \mathbf{E}_k^{\alpha}\mathbf{P}_k^{\beta}:\nabla_{\beta}\mathbf{v}_k^{\alpha}\right\}$$

$$-\sum_{k=0}^{N}\left\{\left(\rho_k\,\frac{d_k\mathbf{m}_k}{dt} + \mathbf{m}_k\sigma_k\right)\cdot\mathbf{H}_k - \mathbf{H}_k^{\alpha}\mathbf{M}_k^{\beta}:\nabla_{\beta}\mathbf{v}_k^{\alpha}\right\}. \qquad (4.25)$$

Finally, from this we have

$$\frac{\partial}{\partial t}\,\rho\varepsilon_e + \nabla_{\beta}\left\{(\mathbf{E}\wedge\mathbf{H})^{\beta} - \sum_{k=0}^{N}\mathbf{v}_k^{\alpha}\,(\mathbf{H}_k^{\alpha}\mathbf{M}_k^{\beta} + \mathbf{E}_k^{\alpha}\mathbf{P}_k^{\beta})\right\}$$

$$= -\sum_{k=0}^{N}\rho_k F_k\cdot\mathbf{v}_k - \sum_{k=0}^{N}\left\{\mathbf{E}_k^{\alpha}\cdot\left(\rho_k\,\frac{d_k\mathbf{p}_k}{dt} + \sigma_k\mathbf{p}_k^{\alpha}\right)\right.$$

$$\left. + \mathbf{H}_k^{\alpha}\cdot\left(\rho_k\,\frac{d_k\mathbf{m}_k}{dt} + \sigma_k\mathbf{m}_k^{\alpha}\right)\right\} - \sum_{k=0}^{N}\mathbf{v}_k^{\alpha}\cdot\nabla_{\beta}\,(\mathbf{E}_k^{\alpha}\mathbf{P}_k^{\beta} + \mathbf{H}_k^{\alpha}\mathbf{M}_k^{\beta}). \qquad (4.26)$$

Comparing the expression obtained with (4.23), we determine

$$\mathbf{S}^{\beta} = (\mathbf{E}\wedge\mathbf{H})^{\beta} - \sum_{k=0}^{N}\mathbf{v}_k^{\alpha}\cdot(\mathbf{H}_k^{\alpha}\mathbf{M}_k^{\beta} + \mathbf{E}_k^{\alpha}\mathbf{P}_k^{\beta}), \qquad (4.27)$$

$$\sigma_k^{e} = \mathbf{E}_k\cdot\left(\rho_k\,\frac{d_k\mathbf{p}_k}{dt} + \sigma_k\mathbf{p}_k\right) + \mathbf{H}_k\cdot\left(\rho_k\,\frac{d_k\mathbf{m}_k}{dt} + \sigma_k\mathbf{m}_k\right), \qquad (4.28)$$

$$\rho_k F_k^{e,\alpha} = \nabla_{\beta}\,(\mathbf{E}_k^{\alpha}\mathbf{P}_k^{\beta} + \mathbf{H}_k^{\alpha}\mathbf{M}_k^{\beta}). \qquad (4.29)$$

We note that the energy flux $\mathbf{S}$ defined in (4.27) includes the energy flux of the electromagnetic field $(\mathbf{E}\wedge\mathbf{H})$ (the Poynting vector) and the energy flux for the interaction of the electromagnetic field with matter during polarization and magnetization. The energy source density σ_k^{e} is determined by the interaction of the field with matter during polarization and magnetization.

The final expression for the force acting on the kth component as viewed from the electromagnetic field (without taking into account the forces arising upon polarization and magnetization, included in the pressure), using (4.9), (4.21), and (4.29) we obtain

$$\rho_k F^{e,\alpha} = \rho_k^e E_k^\alpha + \mu_* (\overset{\text{P}'}{j_k} \wedge H)^\alpha - \varepsilon_* (\overset{\text{M}'}{j_k} \wedge E)^\alpha + P_k^\beta \cdot \nabla_\beta E_k^\alpha + M_k^\beta \cdot \nabla_\beta H_k^\alpha.$$

(4.30)

In conclusion, we give the expressions used later for σ_k^e, $\rho_k F_k^e$:

$$\sigma_k^e = \rho_k \frac{d_k}{dt} (E_k \cdot p_k + H_k \cdot m_k) - \sum_{k=0}^N \left\{ \rho_k P_k \frac{d_k E_k}{dt} - \sigma_k p_k E_k \right\}$$

$$- \sum_{k=0}^N \left\{ \rho_k m_k \cdot \frac{d_k H_k}{dt} - \sigma_k m_k \cdot H_k \right\},$$

(4.31)

$$\rho_k F_k^{e,\alpha} = \rho_k^e E_k^\alpha + \mu_* (\overset{\text{P}'}{j_k} \wedge H)^\alpha - \varepsilon_* (\overset{\text{M}'}{j_k} \wedge E)^\alpha$$
$$+ P_k^\beta \nabla_\alpha E_k^\beta - (P_k \wedge \text{rot } E_k)^\alpha + M_k^\beta \nabla_\alpha H_k^\beta - (M_k \wedge \text{rot } H_k)^\alpha.$$

(4.32)

5. Internal Energy Balance

Let us write the law of conservation of total energy for the kth component in integral form

$$\frac{d}{dt} \int_R \rho_k \left(u_k + \frac{1}{2} v_k \cdot v_k \right) dv + \int_{\partial R} \rho_k \left(u_k + \frac{1}{2} v_k \cdot v_k \right) (v_k \cdot n) \, dA$$

$$= \int_R (\rho_k F_k + \rho_k F_k^e + \rho_k \pi_k) \cdot v_k dv + \int_{\partial R} (T_k \cdot n) \cdot v_k dA + \int_R \lambda_k : \omega_k dv$$

$$- \int_{\partial R} q_k \cdot n dA + \int_R \rho_k r_k dv + \int_R \sigma_k^e dv + \int_R \sigma_k G_k dv + \int_R \psi_k dv,$$

(5.1)

where

$$\lambda_k^{\alpha\beta} : \omega_k^{\alpha\beta} = \lambda_k^{\alpha\beta} \omega_k^{\beta\alpha}, \quad \omega_k^{\alpha\beta} \equiv \frac{1}{2} \left(\frac{\partial v_k^\alpha}{\partial x_\beta} - \frac{\partial v_k^\beta}{\partial x_\alpha} \right) \equiv \frac{1}{2} \{ (\nabla v)_{\beta\alpha} - (\nabla v)_{\alpha\beta} \}.$$

(5.2)

Here u_k is the specific internal energy of the kth component; ω_k is the rotation tensor connected with the kth component; q_k is the heat flux density for the heat supplied to the kth component (i.e., directed into the volume R); $\rho_k r_k$ is the density of the external volume heat sources for the kth component; $\tau_k G_k$ is the density of the internal heat sources for the kth component due to the motion of the mass source; ψ_k is the heat source density for the kth component arising as a result of interaction between the kth component and the rest of the components of the mixture. The right-hand side of expression (5.1) includes the mechanical work done by the volume and surface forces (external forces relative to the kth component), contact and volume supply of heat to the kth component, and influx of heat to the kth component as a result of nonthermal forms of energy, which are not connected with mechanical work. The first three terms on the right-hand side describe the work done by the external forces (volume and surface forces). We note that the force π_k acting on the kth component as viewed from the rest of the components of the mixture is an external force relative to the kth component. The third term on the right-hand side of (5.1) corresponds to the work done by the internal surface forces. The rest of the terms on the right-hand side of (5.1) have the following meaning: the fourth and fifth terms correspond to contact and volume heat supply, the sixth term corresponds to heat arising as a result of interaction between the electromagnetic field and matter upon polarization and magnetization, the seventh term corresponds to heat due to motion of the mass source, and, finally, the eighth term corresponds to heat influx as a result of interaction between components of the mixture (for example, intermolecular interactions of the components).

Taking into account (2.3), we derive the differential equation for internal energy balance for the kth component from (5.1) as

$$\rho_k \frac{d_k u_k}{dt} = \sigma_k \left\{ \mathbf{G}_k - u_k - (\mathbf{J}_k - \mathbf{v}_k) \cdot \mathbf{v}_k - \frac{1}{2} v_k \cdot v_k \right\} + \psi_k$$

$$+ \lambda_k : \omega_k + \rho_k r_k - \operatorname{div} \mathbf{q}_k + \mathbf{T}_k : \nabla \mathbf{v}_k + \sigma_k^e. \tag{5.3}$$

We use the law of conservation of total energy for the mixture in the form

$$\frac{d}{dt} \sum_{k=0}^{N} \int_{R} \rho_k \left(u_k + \frac{1}{2} \mathbf{v}_k \cdot \mathbf{v}_k \right) dv + \sum_{k=0}^{N} \int_{\partial R} \rho_k \left(u_k + \frac{1}{2} \mathbf{v}_k \cdot \mathbf{v}_k \right) (\mathbf{v}_k \cdot \mathbf{n}) dA$$

$$= \sum_{k=0}^{N} \int_{R} (\rho_k \mathbf{F}_k + \rho_k \mathbf{F}_k^e) \cdot \mathbf{v}_k dv + \sum_{k=0}^{N} \int_{\partial R} (\mathbf{T}_k \cdot \mathbf{n}) \cdot \mathbf{v}_k dA$$

$$- \sum_{k=0}^{N} \int_{\partial R} \mathbf{q}_k \cdot \mathbf{n} dA + \sum_{k=0}^{N} \int_{R} \rho_k r_k dv. \tag{5.4}$$

In contrast to (5.1), we do not include the mechanical work done by volume and surface internal forces ($\rho_k \pi_k \cdot \mathbf{v}_k$ and $\lambda_k : \omega_k$), the heat arising upon motion of the mass sources ($\sigma_k G_k$), and the heat connected with interaction between components (ψ_k) on the right-hand side of (5.4). Obviously, by postulating the law of conservation of total energy in the form (5.4), we essentially choose the model for the multicomponent mixture. Summing (5.1) over all components and comparing with (5.4), we derive

$$\sum_{k=0}^{N} \int_R (\rho_k \pi_k \cdot \mathbf{v}_k + \lambda_k : \omega_k + \sigma_k G_k + \psi_k)\, dv = 0. \tag{5.5}$$

In differential form, this equation has the form

$$\sum_{k=0}^{N} (\rho_k \pi_k \cdot \mathbf{v}_k + \lambda_k : \omega_k + \sigma_k G_k + \psi_k) = 0. \tag{5.6}$$

The expression obtained represents the condition which should be satisfied by the internal interactions in the mixture. This expression determines the model of the multicomponent mixture [ultimately, it is equivalent to (5.4)].

In conclusion, let us derive the internal energy balance equation for the mixture. Using (2.5), we write the auxiliary relationship

$$\sum_{k=0}^{N} \rho_k \pi_k \cdot \mathbf{v}_k \equiv \sum_{k=0}^{N} \rho_k \pi_k \cdot \mathbf{w}_k + \sum_{k=0}^{N} \rho_k \pi_k \cdot \mathbf{v}_k = \sum_{k=0}^{N} \rho_k \pi_k \cdot \mathbf{v}_k$$

$$- \sum_{k=0}^{N} \sigma_k \mathbf{J}_k \cdot (\mathbf{v}_k - \mathbf{w}_k). \tag{5.7}$$

Summing (5.3) over all components, using (1.9), (1.13), (5.6), and (5.7), we obtain the desired equation

$$\rho \frac{d}{dt} \sum_{k=0}^{N} (c_k u_k) + \operatorname{div} \sum_{k=0}^{N} (\rho_k u_k \mathbf{w}_k + \mathbf{q}_k) = \sum_{k=0}^{N} \rho_k r_k + \sum_{k=0}^{N} \mathbf{T}_k : \nabla \mathbf{v}_k$$

$$+ \sum_{k=0}^{N} \sigma_k^e - \sum_{k=0}^{N} \rho_k \pi_k \cdot \mathbf{w}_k + \sum_{k=0}^{N} \sigma_k \left\{ \frac{1}{2} \mathbf{w}_k^2 - \mathbf{w}_k \cdot (\mathbf{J}_k - \mathbf{v}) \right\} \equiv \sigma_u, \tag{5.8}$$

where σ_u is the internal energy source density for the mixture.

6. Inequality for the Entropy of the Mixture Components

Suppose that the Clausius–Duhem inequality is satisfied for the kth component (see [52]):

$$-\frac{d}{dt} \int_R \rho_k s_k dv + \int_{\partial R} \rho_k s_k (\mathbf{v}_k \cdot \mathbf{n}) dA - \int_R \frac{\rho_k r_k}{T_k} dv + \int_{\partial R} \frac{1}{T_k} \mathbf{q}_k \cdot \mathbf{n} dA \geqslant 0, \quad (6.1)$$

where s_k is the specific entropy for the kth component; $T_k > 0$ is the temperature of the kth component.

From (6.1) we derive the differential form of the Clausius–Duhem inequality for the kth component:

$$-\frac{\partial}{\partial t} (\rho_k s_k) + \mathrm{div} (\rho_k s_k \mathbf{v}_k) - \frac{\rho_k r_k}{T_k} + \mathrm{div} \left(\frac{1}{T_k} \mathbf{q}_k \right) \geqslant 0. \quad (6.2)$$

We rewrite it in the form

$$\sigma_k^s \equiv \rho_k \frac{d_k s_k}{dt} + \sigma_k s_k - \frac{\rho_k r_k}{T_k} + \mathrm{div} \left(\frac{1}{T_k} \cdot \mathbf{q}_k \right) \geqslant 0, \quad (6.3)$$

where σ_k^s is the entropy source density of the kth component.

Eliminating r_k using (5.3), we derive (if $r_k \equiv 0$, then we eliminate $\mathrm{div}\,\mathbf{q}_k$)

$$T_k \sigma_k^s \equiv \rho_k T_k \frac{d_k s_k}{dt} - \rho_k \frac{d_k u_k}{dt} + T_k \mathbf{q}_k \cdot \nabla \left(\frac{1}{T_k} \right) + \mathbf{T}_k : \nabla \mathbf{v}_k + \psi_k$$

$$+ \lambda_k : \omega_k + \sigma_k \left\{ T_k s_k + G_k - u_k - (\mathbf{J}_k - \mathbf{v}_k) \cdot \mathbf{v}_k - \frac{1}{2} \mathbf{v}_k \cdot \mathbf{v}_k \right\} + \sigma_k^e \geqslant 0. \quad (6.4)$$

7. The Gibbs Relation and the Chemical Potential

We represent the stress tensor for the kth component in the form

$$T_k^{\alpha\beta} = -p_k \delta_{\alpha\beta} + \mathbf{T}_k^{v,\alpha\beta}, \quad (7.1)$$

where p_k is the partial pressure of the kth component and $\mathbf{T}_k^v$ is the viscous stress tensor for the kth component.

Redefine the specific internal energy of the kth component, including the interaction between the electromagnetic field and matter upon polarization and magnetization:

$$u_k^* = u_k - \mathbf{p}_k \cdot \mathbf{E}_k - \mathbf{m}_k \cdot \mathbf{H}_k. \tag{7.2}$$

Next, introduce the specific Helmholtz free energy for the kth component:

$$A_k = u_k^* - T_k s_k. \tag{7.3}$$

It is easy to obtain the relationship

$$\frac{d_k A_k}{dt} + s_k \frac{d_k T_k}{dt} = \frac{du_k^*}{dt} - T_k \frac{d_k s_k}{dt}. \tag{7.4}$$

Substituting the expression obtained into (6.4), taking into account (1.6), (4.31), and (7.2), we derive

$$T_k \sigma_k^s \equiv \rho_k \left\{ - \frac{d_k A_k}{dt} - s_k \frac{d_k T_k}{dt} - p_k \frac{d_k}{dt} \left(\frac{1}{\rho_k} \right) - \mathbf{p}_k \frac{d_k \mathbf{E}_k}{dt} \right.$$

$$\left. - \mathbf{m}_k \frac{d_k \mathbf{H}_k}{dt} \right\} + \sigma_k \left\{ - A_k - \frac{p_k}{\rho_k} + G_k - (\mathbf{J}_k - \mathbf{v}_k) \cdot \mathbf{v}_k - \frac{1}{2} \mathbf{v}_k \cdot \mathbf{v}_k \right\}$$

$$+ T_k \mathbf{q}_k \cdot \nabla \left(\frac{1}{T_k} \right) + \psi_k + \lambda_k : \omega_k + \mathbf{T}_k^v : \nabla \mathbf{v}_k \geqslant 0. \tag{7.5}$$

Let us define the specific chemical potential of the kth component characterizing the change in energy of the kth component as a result of change in mass:

$$\mu_k = A_k + \frac{p_k}{\rho_k} \equiv u_k - T_k s_k + \frac{p_k}{\rho_k} - \mathbf{p}_k \cdot \mathbf{E}_k - \mathbf{m}_k \cdot \mathbf{H}_k. \tag{7.6}$$

Here we assume that the specific free energy of the kth component depends on the specific volumes $1/\rho_n$, temperatures T_n, the electric field intensity $\mathbf{E}_n$, and the magnetic field intensity $\mathbf{H}_n$ for all the components of the mixture $(n = 0, ..., N)$:

$$A_k = A_k(1/\rho_0, \ldots, 1/\rho_N; T_0, \ldots T_N; \mathbf{E}_0, \ldots, \mathbf{E}_N; \mathbf{H}_0, \ldots, \mathbf{H}_N). \tag{7.7}$$

In the case where we choose the quantities $1/\rho_n$, T_n, $\mathbf{E}_n$, $\mathbf{H}_n$ $(n = 0, ..., N)$ as the parameters describing the behavior of the mixture, the dependence in (7.7) is very general and does not impose any restrictions on the model for the multicomponent mixture.

Assuming that the function A_k is continuously differentiable with respect to all the arguments, using (7.5) we derive

$$T_k \sigma_k^s \equiv \rho_k \left\{ - \left(\frac{\partial A_k}{\partial T_k} + s_k \right) \frac{d_k T_k}{dt} - \left(\frac{\partial A_k}{\partial 1/\rho_k} + p_k \right) \frac{d_k}{dt} \frac{1}{\rho_k} - \right.$$

$$-\left(\frac{\partial A_k}{\partial E_k^\alpha} + p_k^\alpha\right) \cdot \frac{d_k E_k^\alpha}{dt} - \left(\frac{\partial A_k}{\partial H_p^\alpha} + m_k^\alpha\right) \cdot \frac{d_k H_k^\alpha}{dt} - \sum_{n \neq k} \frac{\partial A_k}{\partial T_n} \frac{d_k T_n}{dt}$$

$$-\sum_{n \neq k} \frac{\partial A_k}{\partial 1/\rho_n} \frac{d_n}{dt}\left(\frac{1}{\rho_n}\right) - \sum_{n \neq k} \frac{\partial A_k}{d E_n^\alpha} \cdot \frac{d_n E_n^\alpha}{dt} - \sum_{n \neq k} \frac{\partial A_k}{\partial H_n^\alpha} \cdot \frac{d_n H_n^\alpha}{dt}\Bigg\}$$

$$+ \sigma_k\left\{-\mu_k + G_k - (\mathbf{J}_k - \mathbf{v}_k) \cdot \mathbf{v}_k - \frac{1}{2}\mathbf{v}_k \cdot \mathbf{v}_k\right\}$$

$$+ T_k \mathbf{q}_k \cdot \nabla\left(\frac{1}{T_k}\right) + \psi_k + \lambda_k : \omega_k + \mathbf{T}_k^v : \nabla \mathbf{v}_k \geqslant 0. \qquad (7.8)$$

All the allowable changes in the quantities $1/\rho_n$, T_n, $\mathbf{E}_n$, $\mathbf{H}_n$ ($n = 0, \ldots, N$) should satisfy this inequality. It is easy to show that the following relationships are necessary and sufficient conditions for satisfying inequality (7.8) [52]:

$$\frac{\partial A_k}{\partial T_k} = -s_k, \quad \frac{\partial A_k}{\partial 1/\rho_k} = -p_k, \quad \frac{\partial A_k}{\partial E_k^\alpha} = -p_k^\alpha,$$

$$\frac{\partial A_k}{\partial H_k^\alpha} = -m_k^\alpha, \quad k = 0, \ldots, N, \qquad (7.9)$$

$$\frac{\partial A_k}{\partial T_n} \equiv \frac{\partial A_k}{\partial 1/\rho_n} \equiv \frac{\partial A_k}{\partial E_n^\alpha} \equiv \frac{\partial A_k}{\partial H_n^\alpha} \equiv 0, \quad n \neq k; \quad k, n = 0, \ldots, N, \qquad (7.10)$$

$$T_k \sigma_k^s \equiv -\sigma_k \mu_k + \sigma_k\left\{G_k - \frac{1}{2}\mathbf{v}_k \cdot \mathbf{v}_k - (\mathbf{J}_k - \mathbf{v}_k) \cdot \mathbf{v}_k\right\}$$

$$+ T_k \mathbf{q}_k \cdot \nabla\left(\frac{1}{T_k}\right) + \psi_k + \lambda_k : \omega_k + \mathbf{T}_k^v : \nabla \mathbf{v}_k \geqslant 0, \quad k = 0, \ldots, N. \qquad (7.11)$$

Thus, dependence (7.7) has the form

$$A_k = A_k\,(1/\rho_k, \, T_k, \, \mathbf{E}_k, \, \mathbf{H}_k). \qquad (7.12)$$

The equalities in (7.9) for a given dependence (7.12) determine the specific entropy, partial pressure, specific dipole moment, and specific magnetic moment of the kth component.

Substituting (7.12) into (7.4) and using (7.9), we obtain the Gibbs relation for the kth component:

$$T_k \frac{d_k s_k}{dt} = \frac{d_k u_k^*}{dt} + p_k \frac{d_k}{dt}\left(\frac{1}{\rho_k}\right) + \mathbf{p}_k \cdot \frac{d_k \mathbf{E}_k}{dt} + \mathbf{m}_k \cdot \frac{d_k \mathbf{H}_k}{dt} \qquad (7.13)$$

or, taking into account (7.2), we have

$$T_k \frac{d_k s_k}{dt} = \frac{d_k u_k}{dt} + p_k \frac{d_k}{dt}\left(\frac{1}{\rho_k}\right) - \mathbf{E}_k \cdot \frac{d_k \mathbf{p}_k}{dt} - \mathbf{H}_k \cdot \frac{d_k \mathbf{m}_k}{dt}. \qquad (7.14)$$

From (7.6), using (7.9), we derive a representation for the chemical potential gradient:

$$\nabla_\beta \mu_k = - s_k \nabla_\beta T_k + \frac{1}{\rho_k} \nabla_\beta p_k - p_k^\alpha \nabla_\beta E_k^\alpha - m_k^\alpha \nabla_\beta H_k^\alpha,$$

$$\mu_k = \mu_k (1/\rho_k, \, T_k, \, \mathbf{E}_k, \, \mathbf{H}_k).$$

(7.15)

We can show that when the temperatures of all the components are the same and there is no motion of the individual components relative to the mixture, the Gibbs relation for the multicomponent mixture follows from (7.14). Assume that

$$T_k = T, \quad \mathbf{w}_k = 0, \quad k = 0, \, \ldots, \, N. \tag{7.16}$$

Next introduce the symbols

$$u = \sum_{k=0}^{N} c_k u_k, \quad s = \sum_{k=0}^{N} c_k s_k, \quad p^e = \sum_{k=0}^{N} p_k, \quad \mathbf{p} = \sum_{k=0}^{N} c_k \mathbf{p}_k, \quad \mathbf{m} = \sum_{k=0}^{N} c_k \mathbf{m}_k,$$

(7.17)

$$\mathbf{E}^* = \mathbf{E} - \mu_0 (\mathbf{H} \wedge \mathbf{v}), \quad \mathbf{H}^* = \mathbf{H} + \varepsilon_0 (\mathbf{E} \wedge \mathbf{v}).$$

Here, u is the specific internal energy of the mixture; s is the specific entropy of the mixture; p^e is the pressure in the mixture; $\mathbf{p}$ is the specific dipole moment of the mixture; $\mathbf{m}$ is the specific magnetic moment of the mixture; $\mathbf{E}^*$ and $\mathbf{H}^*$ are the effective electric and magnetic field intensities, respectively (i.e., measured in a system relative to which the mixture is at rest).

Summing (7.14) over all components, taking into account (1.6), (7.6), (7.9), (7.12), (7.16), and (7.17), we obtain the familiar Gibbs relation for a multicomponent mixture (see, for example, [11]):

$$T \frac{ds}{dt} = \frac{du}{dt} + p^e \frac{d}{dt} \left(\frac{1}{\rho} \right) - \mathbf{E}^* \cdot \frac{d\mathbf{p}}{dt} - \mathbf{H}^* \cdot \frac{d\mathbf{m}}{dt} - \sum_{k=0}^{N} \mu_k \frac{dc_k}{dt}. \quad (7.18)$$

8. Entropy Balance Equation for the Mixture

To obtain the entropy balance equation for the mixture, we sum (6.3) over all components, taking into account (7.11):

$$\rho \frac{d}{dt} \sum_{k=0}^{N} c_k s_k + \operatorname{div} \sum_{k=0}^{N} \left(\rho_k s_k \mathbf{w}_k + \frac{\mathbf{q}_k}{T_k} \right) - \sum_{k=0}^{N} \frac{\rho_k r_k}{T_k}$$

$$= - \sum_{k=0}^{N} \frac{1}{T_k} \sigma_k \left\{ \mu_k - G_k + (\mathbf{J}_k - \mathbf{v}_k) \cdot \mathbf{v}_k + \frac{1}{2} \mathbf{v}_k \cdot \mathbf{v}_k \right\}$$

$$+ \sum_{k=0}^{N} \mathbf{q}_k \cdot \nabla \left(\frac{1}{T_k} \right) + \sum_{k=0}^{N} \frac{1}{T_k} \mathbf{T}_k^v : \nabla \mathbf{v}_k + \sum_{k=0}^{N} \frac{1}{T_k} (\psi_k + \lambda_k : \omega_k) \equiv \sigma^s \geqslant 0. \quad (8.1)$$

Taking into account (5.6) and (5.7), this expression takes on the form

$$\rho \frac{d}{dt} \sum_{k=0}^{N} c_k s_k + \operatorname{div} \sum_{k=0}^{N} \left(\rho_k s_k \mathbf{w}_k + \mathbf{q}_k \frac{1}{T_k} \right) - \sum_{k=0}^{N} \frac{\rho_k r_k}{T_k}$$

$$= - \sum_{k=0}^{N} \frac{\sigma_k}{T_k} \left\{ \mu_k - \frac{1}{2} \mathbf{w}_k^2 + \mathbf{w}_k \cdot (\mathbf{J}_k - \mathbf{v}) \right\} - \sum_{k=0}^{N} \frac{1}{T_k} \rho_k \pi_k \cdot \mathbf{w}_k$$

$$+ \sum_{k=0}^{N} \mathbf{q}_k \cdot \nabla \left(\frac{1}{T_k} \right) + \sum_{k=0}^{N} \frac{1}{T_k} \mathbf{T}_k^v : \nabla_k \equiv \sigma^s \geqslant 0. \tag{8.2}$$

The relationship obtained in the case where the temperatures of all the components are the same allows us to derive an equation for determining the temperature of the mixture. Set

$$T_k = T, \quad k = 0, \ldots, N. \tag{8.3}$$

From (7.3) and (7.9) it follows that

$$T \frac{\partial s_k}{\partial T} = \frac{\partial u_k}{\partial T} = c_{v,k}, \quad \left(\frac{\partial u_k}{\partial T} \equiv - T \frac{\partial^2 A_k}{\partial T^2} \right) \quad (k = 0, \ldots, N), \tag{8.4}$$

where $c_{v,k}$ is the specific heat for the kth component at constant volume.

Next write the reciprocal Maxwell relations, which are easily derived from (7.9) (see, for example, [29]):

$$\frac{\partial s_k}{\partial 1/\rho_k} = \frac{\partial p_k}{\partial T}, \quad \frac{\partial s_k}{\partial E_k^\alpha} = \frac{\partial p_k^\alpha}{\partial T}, \quad \frac{\partial s_k}{\partial H_k^\alpha} = \frac{\partial m_k^\alpha}{\partial T}, \quad k = 0, \ldots, N. \tag{8.5}$$

Then we introduce the specific enthalpy of the kth component:

$$h_k = u_k^\cdot + \frac{p_k}{\rho_k} \equiv \mu_k + T s_k, \quad k = 0, \ldots, N. \tag{8.6}$$

Excluding the quantity $\rho_k \pi_k$ from (8.2) using (2.3), taking into account expressions (1.8), (1.9), (4.33), and (7.15), we obtain the equation for determining the temperature of the mixture

$$\rho \left(\sum_{k=0}^{N} c_k c_{v,k} \right) \frac{dT}{dt} + \operatorname{div} \sum_{k=0}^{N} (\mathbf{q}_k + \mathbf{i}_k h_k) = \sum_{k=0}^{N} \rho_k r_k - \sum_{k=0}^{N} \sigma_k \frac{1}{2} \mathbf{w}_k^2$$

$$+ \sum_{k=0}^{N} (e_k \mathbf{i}_k - \rho \sum_{k=0}^{N} h_k \frac{dc_k}{dt} + \sum_{k=0}^{N} \mathbf{w}_k \cdot \left\{ \mu_* (\mathbf{j}_k^{\mathbf{P}} \wedge \mathbf{H}) - \varepsilon_* (\mathbf{j}_k^{\mathbf{M'}} \wedge \mathbf{E}) - \right.$$

$$- \mathbf{p}_k \wedge \text{rot } \mathbf{E}_k - \mathbf{M}_k \wedge \text{rot } \mathbf{H}_k + \text{div } \mathbf{T}_k^v - \rho_k \frac{d_k \mathbf{v}_k}{dt} \Big\} + \sum_{k=0}^{N} \mathbf{T}_k^v : \nabla \mathbf{v}_k$$

$$- \rho \sum_{k=0}^{N} c_k \left(T \frac{\partial p_k}{\partial T} \frac{d}{dt} \frac{1}{\rho_k} + T \frac{\partial p_k}{\partial T} \cdot \frac{d \mathbf{E}_k}{dt} + T \frac{\partial \mathbf{m}_k}{\partial T} \frac{\partial \mathbf{H}_k}{dt} \right) + \sum_{k=0}^{N} \mathbf{F}_k \cdot \mathbf{i}_k.$$

$$(8.7)$$

9. Description of the Behavior of Multicomponent Mixtures

Let us use the balance equations obtained in the preceding sections to describe the behavior of multicomponent mixtures. Recall the basic assumptions made in deriving the balance equations. The law of conservation of mass, (1.9), is satisfied for the mixture as a whole. The density of the sum of internal forces is balanced by the sum of the momenta arising upon motion of the mass sources, (2.5). Maxwell's equations for the mixture may be written in the form (4.1)-(4.4); the sources of the fields, i.e., the charge densities, have the form (4.5), (4.6), (4.9), (4.10), and (4.12). The Lorentz forces acting on the charges have the form (4.19); the action of the electromagnetic forces on the mixture is determined by the expression (4.32). The law of conservation of energy for the mixture has the form (5.4) or, equivalently, condition (5.6) is satisfied. Finally, the Clausius–Duhem inequality is satisfied for each component of the mixture, (6.2).

The model of the multicomponent mixture constructed on the basis of the assumptions listed is very general. Next we consider some possible simplifications of this model. Assume that the temperature of all the components in the mixture is the same:

$$T_k = T, \quad k = 0, \ldots N. \tag{9.1}$$

Now isolate one component of the mixture, for example the zeroth (0th) component, which we will call the solvent. In order to describe the behavior of the mixture, choose the following parameters: the density of the mixture ρ, the velocity of the particle in the mixture $\mathbf{v}$, the density of the mixture components ρ_k ($k = 1, \ldots, N$), the velocities of the mixture components $\mathbf{v}_k$ ($k = 1, \ldots, N$), the electric field intensity $\mathbf{E}$, the magnetic field intensity $\mathbf{H}$, and the temperature of the mixture T. To determine these quantities, we have Eqs. (1.3), (1.10), (2.3), (2.4), (4.15)-(4.18), and (8.7), which for convenience we represent in the following form:

$$\frac{d\rho}{dt} + \rho \, \text{div } \mathbf{v} = 0, \tag{9.2}$$

$$\frac{\partial \rho_k}{\partial t} + \mathrm{div}\,(\rho_k \mathbf{v}_k) = o_k, \quad k = 1, \ldots, N, \tag{9.3}$$

$$\rho\,\frac{d\mathbf{v}}{dt} = \mathrm{div}\,\sum_{k=0}^{N} (\mathbf{T}_k - \rho_k \mathbf{w}_k \mathbf{w}_k) + \sum_{k=0}^{N} \rho_k \mathbf{F}_k + \sum_{k=0}^{N} \rho_k \mathbf{F}_k^e, \tag{9.4}$$

$$\rho_k\,\frac{d_k \mathbf{v}_k}{dt} = \mathrm{div}\,\mathbf{T}_k + \sigma_k\,(\mathbf{J}_k - \mathbf{v}_k) + \rho_k\,(\mathbf{F}_k + \mathbf{F}_k^e + \boldsymbol{\pi}_k), \quad k = 1, \ldots, N, \tag{9.5}$$

$$\mathrm{rot}\left(\mathbf{H} - \sum_{k=0}^{N} \mathbf{P}_k \wedge \mathbf{v}_k\right) = \frac{\partial}{\partial t}\left(\varepsilon_0 \mathbf{E} + \sum_{k=0}^{N} \mathbf{P}_k\right) + \sum_{k=0}^{N} \rho_k e_k \mathbf{v}_k, \tag{9.6}$$

$$\mathrm{rot}\left(\mathbf{E} + \sum_{k=0}^{N} \mathbf{M}_k \wedge \mathbf{v}_k\right) = -\frac{\partial}{\partial t}\left(\mu_0 \mathbf{H} + \sum_{k=0}^{N} \mathbf{M}_k\right), \tag{9.7}$$

$$\mathrm{div}\left(\mu_0 \mathbf{H} + \sum_{k=0}^{N} \mathbf{M}_k\right) = 0, \quad \mathrm{div}\left(\varepsilon_0 \mathbf{E} + \sum_{k=0}^{N} \mathbf{P}_k\right) = 0, \tag{9.8}$$

$$\rho\left(\sum_{k=0}^{N} c_k c_{v,k}\right)\frac{dT}{dt} + \mathrm{div}\,\sum_{k=0}^{N} (\mathbf{q}_k + \mathbf{i}_k h_k) = \sum_{k=0}^{N} \rho_k r_k$$

$$-\sum_{k=0}^{N} \sigma_k \frac{1}{2}\,\mathbf{w}_k^2 + \sum_{k=0}^{N} e_k \mathbf{i}_k \cdot \mathbf{E}_k - \rho \sum_{k=0}^{N} h_k \frac{dc_k}{dt} + \sum_{k=0}^{N} \mathbf{F}_k \cdot \mathbf{i}_k + \sum_{k=0}^{N} \mathbf{T} : \nabla \mathbf{v}_k$$

$$+ \sum_{k=0}^{N} \mathbf{w}_k \cdot \left\{ \mu_*\,(\mathbf{j}_k^{\mathbf{P}'} \wedge \mathbf{H}) - \varepsilon_*\,(\mathbf{j}_k^{\mathbf{M}'} \wedge \mathbf{E}) \right.$$

$$\left. - \mathbf{P}_k \wedge \mathrm{rot}\,\mathbf{E}_k - \mathbf{M}_k \wedge \mathrm{rot}\,\mathbf{H}_k + \mathrm{div}\,\mathbf{T}_k^v - \rho_k \frac{d_k \mathbf{v}_k}{dt} \right\}$$

$$- \rho \sum_{k=0}^{N} c_k\left(\mathbf{T}\,\frac{\partial p_k}{\partial T}\,\frac{d}{dt}\,\frac{1}{\rho_k} + T\,\frac{\partial p_k}{\partial T} \cdot \frac{dE_k}{dt} + T\,\frac{\partial m_k}{\partial T} \cdot \frac{dH_k}{dt}\right). \tag{9.9}$$

The quantities $\mathbf{P}_k$, $\mathbf{M}_k$, $\mathbf{j}_k^{\mathbf{P}'}$, $\mathbf{j}_k^{\mathbf{M}'}$, $\mathbf{F}_k^e$ entering into (9.4)-(9.9) are specified with the aid of expressions (4.7), (4.12), and (4.32). Finally, the system (9.2)-(9.9) is not closed. For closure, we need to specify $(N + 1)$ equations of state of the type (7.12) for the quantities A_k, allowing us to determine $c_{v,k}$, $\partial p_k/\partial T$, $\partial m_k/\partial T$, $\partial p_k/\partial T$, p_k, $\mathbf{P}_k$, m_k, p_k, $\mathbf{M}_k$, $\mathbf{j}_k^{\mathbf{P}'}$, $\mathbf{j}_k^{\mathbf{M}'}$; $2(N + 1)$ quantities e_k, h_k characterizing the mixture; $3(N + 1)$ components of the external forces F_k; $13N + 10$ scalar defining relations for $\mathbf{T}_k^v$ ($k = 0$, ..., N), σ_k, π_k, $\mathbf{J}_k$ ($k = 1, \ldots, N$), $\sum_{k=0}^{N} \rho_k r_k$, $\sum_{k=0}^{N} \mathbf{q}_k$.

The considered model for a multicomponent mixture is still very complex. In the following, we will develop simpler models for multicomponent mixtures which are used in concrete problems. Therefore, in this chapter we do not consider methods for constructing the defining relations.

10. Simplified Equations of Motion and the Entropy Balance Equation

To show how the description of a multicomponent mixture may be substantially simplified, consider the model for a multicomponent mixture presented in Section 9. Assume that in the equations of motion for the individual components (9.5) [or (2.3)], we may neglect the inertial force connected with motion of the component relative to the mixture, and the momentum transport by the mass source upon motion relative to the mixture. In other words, in (9.5) we discard terms of the type

$$\rho_k \frac{d_k w_k}{dt} + \rho_k w_k \cdot \nabla v + \sigma_k w_k. \tag{10.1}$$

Then (9.5) [or (2.3)] may be considered as a definition for the internal force:

$$\rho_k \pi_k = - \rho_k (F_k + F_k^e) - \sigma_k (J_k - v) - \text{div } T_k + \rho_k \frac{dv}{dt}, \quad k = 0, \ldots, N. \tag{10.2}$$

Summing (10.2) over all components, taking into account (2.5), we obtain the equation of motion for the mixture as a whole [cf. (9.4) or (2.4)]:

$$\rho \frac{dv}{dt} = \text{div} \sum_{k=0}^{N} T_k + \sum_{k=0}^{N} \rho_k (F_k + F_k^e). \tag{10.3}$$

Equation (10.2) may be used to simplify the entropy balance equation. Specifically, having eliminated $\rho_k \pi_k$ from (8.2) and taking account of (10.2), (1.8), (1.9), (4.32), and (7.15), we obtain the entropy balance equation for the mixture by a derivation analogous to (8.7):

$$\rho_k \frac{d}{dt} \sum_{k=0}^{N} c_k s_k + \text{div} \sum_{k=0}^{N} \left(i_k s_k + \frac{q_k}{T_k} \right) - \sum_{k=0}^{N} \frac{\rho_k r_k}{T_k}$$

$$= - \sum_{k=0}^{N} \frac{\sigma_k}{T_k} \left(\mu_k - \frac{1}{2} w_k^2 \right) + \sum_{k=0}^{N} (q_n + i_k h_k) \cdot \nabla \left(\frac{1}{T_k} \right)$$

$$+ \sum_{k=0}^{N} \frac{1}{T_k} T_k^v : \nabla v + \sum_{k=0}^{N} i_k \cdot \frac{1}{T_k} \cdot \left\{ - T_k \nabla \left(\frac{\mu_k}{T_k} \right) + F_k + e_k E_k \right\}$$

$$+ \sum_{k=0}^{N} \frac{1}{T_k} w_k \{ \mu_0 (j_k^{P'} \wedge H) - \varepsilon_0 (j_k^{M'} \wedge E) - P_k \wedge \text{rot } E_k - M_k \wedge \text{rot } H_k \}$$

$$+ \sum_{k=0}^{N} \text{div} \left(\frac{1}{T_k} \right) w^\alpha T_k^{v,\alpha\beta}) - \sum_{k=0}^{N} w_k^\sigma \cdot T_k^{v,\alpha\beta} \nabla_\beta \frac{1}{T_k} \right) \equiv \sigma^s \geqslant 0. \tag{10.4}$$

In the case where the temperatures of all the components are the same ($T_k = T$), the equation for determining the temperature, analogous to (8.7), has the form

$$\rho \left(\sum_{k=0}^{N} c_{v,k} c_k \right) \frac{dT}{dt} + \text{div} \sum_{k=0}^{N} (\mathbf{q}_k + \mathbf{i}_k h_k) = \sum_{k=0}^{N} \rho_k r_k$$

$$+ \sum_{k=0}^{N} e_k \mathbf{i}_k \cdot \mathbf{E}_k + \sum_{k=0}^{N} \mathbf{T}_k^v : \nabla \mathbf{v} + \sum_{k=0}^{N} \mathbf{F}_k \cdot \mathbf{i}_k - \sum_{k=0}^{N} \sigma_k \frac{1}{2} \mathbf{w}_k^2$$

$$- \rho \sum_{k=0}^{N} h_k \frac{dc_k}{dt} + \sum_{k=0}^{N} \mathbf{w}_k \cdot \{ \mu_* (\mathbf{j}_k^{P'} \wedge \mathbf{H}) - \varepsilon_* (\mathbf{j}_k^{M'} \wedge \mathbf{E})$$

$$- \mathbf{P}_k \wedge \text{rot} \mathbf{E}_k - \mathbf{M}_k \wedge \text{rot} \mathbf{H} \} - \rho \sum_{k=0}^{N} c_k \left\{ T \frac{\partial \rho_k}{\partial T} \frac{d}{dt} \frac{1}{\rho_k} \right.$$

$$\left. + T \frac{\partial \mathbf{p}_k}{\partial T} \cdot \frac{d\mathbf{E}_k}{dt} + T \frac{\partial \mathbf{m}_k}{\partial T} \frac{d\mathbf{H}_k}{dt} \right\} + \text{div} \sum_{k=0}^{N} \mathbf{w}_k^\alpha \cdot \mathbf{T}_k^{v,\alpha\beta}. \qquad (10.5)$$

If we replace (10.1) by the expression

$$\rho_k \frac{d_k \mathbf{w}_k}{dt} + \sigma_k \mathbf{w}_k, \qquad (10.1')$$

then (10.2) takes on the form

$$\rho_k \pi_k = - \rho_k (\mathbf{F}_k + \mathbf{F}_k^e) - \sigma_k (\mathbf{J}_k - \mathbf{v}) - \text{div} \mathbf{T}_k + \rho_k \frac{d\mathbf{v}}{dt} + \rho_k \mathbf{w}_k \cdot \nabla \mathbf{v}. \qquad (10.2')$$

Expression (10.3) remains as before; the $\sum_{k=0}^{N} (1/T_k) \mathbf{T}_k^v : \nabla \mathbf{v}$ in (10.4) and (10.5) are replaced by the following: $\sum_{k=0}^{N} (1/T_k)(\mathbf{T}_k^v - \rho_k \mathbf{w}_k \mathbf{w}_k) : \nabla \mathbf{v}$.

In the model considered there are no equations for determining the quantities $\mathbf{v}_k = \mathbf{v} + \mathbf{w}_k$; they are replaced by the definitions of the internal forces. Therefore, in order to describe the behavior of the mixture, we choose the $N + 11$ quantities ρ, $\mathbf{v}$, ρ_k ($k = 1, ..., N$), $\mathbf{E}$, $\mathbf{H}$, T. For their determination we have Eqs. (9.2), (9.3), (9.6)-(9.8), (10.3), and (10.5). Furthermore, we need to specify the $N + 1$ equations of state for the A_k; the $2(N + 1)$ quantities e_k, h_k; the $3(N + 1)$ components of the external forces $\mathbf{F}_k$; and the $N + 10$ scalar defining relations for σ_k ($k = 1, ..., N$),

$$\sum_{k=0}^{N} \mathbf{T}_k^v, \sum_{k=0}^{N} \rho_k r_k, \sum_{k=0}^{N} (\mathbf{q}_k + \mathbf{i}_k h_k - \mathbf{w}_k^\alpha \cdot \mathbf{T}_k^{v,\alpha\beta}).$$

11. Basic Model for a Multicomponent Mixture

In this section, we will consider the model for a multicomponent mixture that is used later. Let the temperature of all the mixture components be the same:

$$T_k = T, \quad k = 0, \ldots, N. \tag{11.1}$$

In the equations of motion we will neglect the inertial terms connected with the motion of the components relative to the mixture and the effect of momentum transport due to motion of the mass sources of the components relative to the mixture, i.e., terms of the type

$$\rho_k \frac{d_k \mathbf{w}_k}{dt} + \rho_k \mathbf{w}_k \cdot \nabla \mathbf{v} + \sigma_k \mathbf{w}_k, \quad k = 0, \ldots, N. \tag{11.2}$$

Let us assume that for all the components in the mixture there are no magnetization or conduction current densities not connected with component motion (the Helmholtz free energy does not depend on the magnetic field intensity):

$$\mathbf{m}_k \equiv -\frac{\partial A_k}{\partial \mathbf{H}_k} \equiv 0, \quad \mathbf{M}_k = 0, \quad \mathbf{j}_k^{\mathbf{M}} = 0, \quad k = 0, \ldots, N. \tag{11.3}$$

We will consider the magnetic field in the mixture to be weak and slowly varying, i.e., we will neglect terms of the type $\mu_* \mathbf{H}$ and $\mu_* (\partial \mathbf{H}/\partial t)$ (but not the terms rot $\mathbf{H}$). Furthermore, in the equation for determining the temperature and the entropy balance equation, we will neglect the heat flux connected with transport of the viscous stress tensor, i.e., the term of the form div $\left(\sum_{k=0}^{N} \mathbf{w}_k^{\alpha} \cdot \mathbf{T}_k^{v,\alpha\beta} \right)$. Introduce the symbols

$$\left. \begin{aligned} \mathbf{T}_\beta^\alpha &= -p_p^e \delta_{\alpha\beta} + \mathbf{T}^v, \quad \mathbf{T}^v = \sum_{k=0}^{N} \mathbf{T}^v, \quad p^e = \sum_{k=0}^{N} p_k, \\ s &= \sum_{k=0}^{N} c_k s_k, \quad \mathbf{q} = \sum_{k=0}^{N} (\mathbf{q}_k + \mathbf{i}_k h_k), \quad \mathbf{J}_s = \frac{1}{T} \mathbf{q}, \\ \rho r &= \sum_{k=0}^{N} \rho_k r_k. \end{aligned} \right\} \tag{11.4}$$

Here $\mathbf{T}$ is the stress tensor for the mixture; p^e is the pressure of the mixture, including the electromagnetic pressure; $\mathbf{T}^v$ is the viscous stress tensor of the mixture; s is the specific entropy of the mixture; $\mathbf{q}$ is the heat flux density supplied to the mixture; ρr is the external heat source density; $\mathbf{J}_e$ is the entropy flux density of the mixture.

As the parameters for describing the behavior of the mixture, choose the following $N + 8$ quantities: the density of the mixture ρ, velocity of the mixture $\mathbf{v}$, concentrations of the components c_k ($k = 1, \ldots, N$), electric field intensity $\mathbf{E}$, and temperature T. In order to determine these quantities for the assumptions made, we have the system of equations

$$\frac{d\rho}{dt} + \rho \operatorname{div} \mathbf{v} = 0, \tag{11.5}$$

$$\rho \frac{dc_k}{dt} + \operatorname{div} \mathbf{i}_k = \sigma_k, \quad k = 1, \ldots, N, \tag{11.6}$$

$$\rho \frac{d\mathbf{v}}{dt} = -\nabla p^e + \operatorname{div} \mathbf{T}^v + \sum_{k=0}^{N} \rho c_k \{\mathbf{F}_k + e_k \mathbf{E} + (\nabla_\alpha \mathbf{T}^\beta) \cdot \mathbf{P}_k^\beta\}, \tag{11.7}$$

$$\frac{\partial}{\partial t} \left(\varepsilon_* \mathbf{E} + \sum_{k=0}^{N} \rho c_k \mathbf{p}_k \right) + \operatorname{rot} \sum_{k=0}^{N} \rho c_k (\mathbf{p}_k \wedge \mathbf{v}_k) + \sum_{k=0}^{N} e_k \mathbf{i}_k + \sum_{k=0}^{N} \rho c_k e_k \mathbf{v} = \mathbf{J}, \tag{11.8}$$

$$\operatorname{div} \left(\varepsilon_* \mathbf{E} + \rho \sum_{k=0}^{N} c_k \mathbf{p}_k \right) = \rho \sum_{k=0}^{N} c_k e_k, \quad \operatorname{rot} \mathbf{E} = 0, \quad \operatorname{div} \mathbf{J} = 0, \tag{11.9}$$

$$\rho \left(\sum_{k=0}^{N} c_k c_{v,k} \right) \frac{dT}{dt} + \operatorname{div} \mathbf{q} = \rho r + \sum_{k=0}^{N} e_k \mathbf{i}_k \cdot \mathbf{E} + \mathbf{T}^v : \nabla \mathbf{v}$$

$$+ \sum_{k=0}^{N} \mathbf{F}_k \cdot \mathbf{i}_k - \sum_{k=0}^{N} \frac{1}{2} \sigma_k \mathbf{w}_k^2 - \rho \sum_{k=0}^{N} h_k \frac{dc_k}{dt}$$

$$- \rho \sum_{k=0}^{N} c_k \left\{ T \frac{\partial p_k}{\partial T} \frac{d}{dt} \left(\frac{1}{\rho c_k} \right) + T \frac{\partial \mathbf{p}_k}{\partial T} \cdot \frac{d\mathbf{E}}{dt} \right\}, \tag{11.10}$$

where $\mathbf{J} \equiv \operatorname{rot} \mathbf{H}$ is the external current density in the circuit.

The entropy balance equation has the form

$$\rho \frac{ds}{dt} + \operatorname{div} \mathbf{J}_s - \frac{1}{T} \rho r = -\sum_{k=0}^{N} \frac{\sigma_k}{T} \mu_k^* + \mathbf{q} \cdot \nabla \left(\frac{1}{T} \right)$$

$$+ \frac{1}{T} \mathbf{T}^v : \nabla \mathbf{v} + \sum_{k=0}^{N} \frac{1}{T} \mathbf{i}_k \cdot \left\{ \mathbf{F}_k + e_k \mathbf{E} - T \nabla \left(\frac{\mu_k}{T} \right) \right\} \equiv \sigma^s \geq 0, \tag{11.11}$$

where

$$\mu_k^* \equiv \mu_k - \frac{1}{2} \mathbf{w}_k^2 \tag{11.12}$$

is the specific chemical potential for a fixed kth component.

The system of equations (11.5)-(11.10) must be supplemented by the equations of state for the A_k ($k = 0, \ldots, N$); the quantities r, e_k, h_k, F_k ($k = 0, \ldots, N$) and the defining relations for i_k ($k = 1, \ldots, N$), $\mathbf{T}^v$, $\mathbf{q}$, σ_k ($k = 1, \ldots, N$) must also be specified.

12. Defining Relations for the Specific Dipole Moment

The equation of state (7.12) allows us to determine how the specific dipole moment depends on the electric field intensity. Consider an isotropic multicomponent mixture. In this case, the Helmholtz free energy does not depend on the direction of the electric field intensity, but depends only on the magnitude of the field intensity [32]:

$$A_k = A_k(T, 1/\rho_k, \mathbf{E}^2), \quad k = 0, \ldots, N. \tag{12.1}$$

The dielectric susceptibility of the kth component is

$$\varkappa_k = -2\rho \frac{1}{\varepsilon_*} \frac{\partial A_k}{\partial (\mathbf{E}^2)}, \quad \varkappa_k = \varkappa_k(T, 1/\rho_k, \mathbf{E}^2) \quad k = 0, \ldots, N. \tag{12.2}$$

Taking into account (7.9) and (4.7), for the polarization vector and the specific dipole moment we have the expressions

$$\mathbf{P}_k = \frac{\varkappa_k \varepsilon_*}{\rho} \mathbf{E}, \quad \mathbf{p}_k = c_k \varkappa_k \varepsilon_* \mathbf{E}, \quad k = 0, \ldots, N. \tag{12.3}$$

The polarization vector of the mixture $\mathbf{P}$ and the dielectric susceptibility of the mixture $\varkappa$ follow as

$$\mathbf{P} = \sum_{k=0}^{N} \mathbf{P}_k, \quad \mathbf{P} = \varkappa \varepsilon_* \mathbf{E}. \tag{12.4}$$

From this, using (12.3), we obtain the relation

$$\varkappa = \sum_{k=0}^{N} c_k \varkappa_k \equiv \varkappa_0 + \sum_{k=0}^{N} (\varkappa_k - \varkappa_0) c_k. \tag{12.5}$$

Analogously, introducing the dielectric constant of the kth component ε_k and the dielectric constant of the mixture ε, we derive

$$\varepsilon = \sum_{k=0}^{N} c_k \varepsilon_k \equiv \varepsilon_0 + \sum_{k=0}^{N} (\varepsilon_k - \varepsilon_0) c_k, \quad \varepsilon_k \equiv 1 + \varkappa_k, \quad \varepsilon \equiv 1 + \varkappa. \tag{12.6}$$

Recall that the pressure in the mixture, p^e, entering into (11.7) includes the pressure arising upon interaction between the electromagnetic field and matter. It is expedient to isolate from this the quantity connected with the electromagnetic field, having included it in the force acting on the mixture. In this case, (11.7) will have the form

$$\rho \frac{dv}{dt} = - \nabla p + \operatorname{div} \mathbf{T}^v + \sum_{k=0}^{N} \rho c_k \{\mathbf{F}_k + e_k \mathbf{E}\} + \rho f^e, \qquad (12.7)$$

where p is the pressure in the mixture in the absence of an electric field; f^e is the specific force acting on the mixture as viewed from the electromagnetic field upon polarization.

Next we determine the quantity ρf^e in the case where the dielectric susceptibility does not depend on the electric field intensity, i.e.,

$$\varkappa_k = \varkappa_k(T, 1/\rho_k), \quad \frac{\partial \varkappa_k}{\partial \mathbf{E}^2} \equiv 0, \quad k = 0, \ldots, N. \qquad (12.8)$$

Introduce the Helmholtz specific free energy for the mixture:

$$A = \sum_{k=0}^{N} c_k A_k. \qquad (12.9)$$

Using (12.2) and (12.5), we obtain

$$\frac{\partial A}{\partial (\mathbf{E}^2)} = - \frac{\varepsilon_* \varkappa}{2\rho}, \quad \frac{\partial \varkappa}{\partial (\mathbf{E}^2)} \equiv 0. \qquad (12.10)$$

Integrating this, we determine

$$A = A^{(0)} - \frac{1}{2} \frac{\varepsilon_* \varkappa}{\rho} \mathbf{E}^2, \quad A^{(0)} = A^{(0)}(T, 1/\rho_0, 1/\rho_1, \ldots, 1/\rho_N). \qquad (12.11)$$

The pressure in the mixture in the absence of an electromagnetic field is

$$p = - \frac{\partial A^{(0)}}{\partial 1/\rho}. \qquad (12.12)$$

Using (7.9), we derive

$$\frac{\partial A}{\partial 1/\rho} = \sum_{k=0}^{N} c_k \frac{\partial A_k}{\partial 1/\rho_k} \cdot \frac{\partial 1/\rho_k}{\partial 1/\rho} = - \sum_{k=0}^{N} p_k = - p^c. \qquad (12.13)$$

Taking into account (12.11) and (12.12), we have

$$p^e = p + \frac{1}{2}\,\varepsilon_*\varkappa E^2 - \frac{1}{2}\,\varepsilon_*\rho\,\frac{\partial\varkappa}{\partial\rho}\,E^2. \tag{12.14}$$

Substituting the relationship obtained into (11.7) and comparing with (12.7), we derive an expression for the force acting on the mixture as viewed from the electromagnetic field upon polarization:

$$\rho f^e = -\frac{1}{2}\,\nabla\,(\varkappa\varepsilon_*)\,E^2 + \frac{1}{2}\,\nabla\left(\rho\,\frac{\partial\varepsilon_*\varkappa}{\partial\rho}\,E^2\right). \tag{12.15}$$

We note that the last term in (12.15) describes the electrostriction forces arising in a liquid dielectric.

13. Linear Onsager Defining Relations

As the defining relations for $\mathbf{T}^v$, $\mathbf{q}$, $\mathbf{i}_k$, σ_k ($k = 1, \ldots, N$), we will use the Onsager relations. Considering that the application of the Onsager relations have been repeatedly described in the literature (see, for example, [10, 11, 14, 38]), we will not give their general form, limiting ourselves to several special cases. We will use the viscous stress tensor in the mixture in the form

$$\mathbf{T}^{v,\alpha\beta} = 2\eta\,(\nabla\mathbf{v})^{s,\alpha\beta} - \left(\frac{2}{3}\,\eta - \eta_v\right)\operatorname{div}\mathbf{v}\cdot\delta_{\alpha\beta}, \tag{13.1}$$

$$\mathbf{T}^{v,\alpha\beta} = \eta\left(\frac{\partial v^\alpha}{\partial x_\beta} + \frac{\partial v^\beta}{\partial x_\alpha}\right) - \left(\frac{2}{3}\,\eta - \eta_v\right)\frac{\partial v^\gamma}{\partial x_\gamma}\,\delta_{\alpha\beta}. \tag{13.2}$$

Here η is the shear viscosity and η_v is the bulk viscosity. We will write the defining relations for the mass sources σ_k using the equations of chemical kinetics (the specific expressions will be given in each special case). We note that the expressions obtained for mass sources using the laws of chemical kinetics may be isolated only when we use the nonlinear Onsager relations (see [3, 38]). We will write the Onsager relations directly only for the quantities $\mathbf{q}$ and $\mathbf{i}_k$ ($k = 1, \ldots, N$) in the case of an isotropic mixture.

Based on expression (11.11) for the entropy source of the mixture, introduce the thermodynamic forces [11, 14]

$$\mathbf{X}_k = \frac{1}{T}\,(\mathbf{F}_k - \mathbf{F}_0) + \frac{1}{T}\,(e_k - e_0)\,\mathbf{E} - \nabla\frac{\mu_k - \mu_0}{T}\,, \quad k = 1, \ldots, N, \tag{13.3}$$

$$\mathbf{X}_q = \nabla\left(\frac{1}{T}\right). \tag{13.4}$$

Here the $\mathbf{X}_k$ are the thermodynamic diffusion forces associated with the diffusion flux density $\mathbf{i}_k$; $\mathbf{X}_q$ is the thermodynamic force associated with the heat flux density. We note that the part of the entropy source density for the mixture, $\sigma_*{}^s$, corresponding to the diffusion process, thermal conductivity, and generation of the conduction current, has the form

$$\sigma_*^s \equiv \sum_{k=0}^{N} \mathbf{i}_k \cdot \mathbf{X}_k + \mathbf{q} \cdot \mathbf{X}_q \geqslant 0. \tag{13.5}$$

We will use the Onsager relations, employing the Curie principle for isotropic mixtures (see [10, 11, 14, 38]), in the form

$$\mathbf{i}_k = \sum_{n=1}^{N} L_{kn}\mathbf{X}_n + L_{kq}\mathbf{X}_q, \quad k = 1, \ldots, N, \tag{13.6}$$

$$\mathbf{q} = \sum_{n=1}^{N} L_{qn}\mathbf{X}_n + L_{qq}\mathbf{X}_q. \tag{13.7}$$

In this case, for the kinetic Onsager coefficients L_{kn} ($k, n = 1, \ldots, N$); L_{kq}, L_{qk} ($k = 1, \ldots, N$), the Onsager–Casimir reciprocal relations are satisfied [11, 14, 38]:

$$L_{kn} = L_{nk}, \quad n, k = 1, \ldots, N. \tag{13.8}$$

Expressions (13.6) and (13.7) are the most general form of the defining relations for the quantities $\mathbf{i}_k$ ($k = 1, \ldots, N$), $\mathbf{q}$.

In the following, we restrict ourselves to a special case of these relations. Consider a model for a multicomponent mixture neglecting the following effects. We will not take into account the diffusion heat flux and, accordingly, the part of the contribution to the diffusion flux from the temperature gradient, having assumed $L_{qn} = L_{nq} = 0$ ($n = 1, \ldots, N$). We note that, in this case, we do not completely neglect the thermal diffusion effect, which will be taken into account in terms connected with the thermodynamic diffusion forces $\mathbf{X}_k$. We will not take into account effects connected with cross-diffusion, requiring $L_{kn} = 0$, $k \neq n$ ($k, n = 1, \ldots, N$). Finally, the Onsager relations take the form

$$\mathbf{i}_k = L_{kk}\mathbf{X}_k \equiv L_{kk}\left\{ \frac{\mathbf{F}_k - \mathbf{F}_0}{T} + \frac{e_k - e_0}{T}\mathbf{E} - \nabla\frac{\mu_k - \mu_0}{T} \right\}, \tag{13.9}$$

$$\mathbf{q} = L_{qq}\nabla\left(\frac{1}{T}\right) \equiv -L_{qq}\frac{1}{T^2}\nabla T. \tag{13.10}$$

Further refinement of the defining relations is possible if we know the dependence of the relative chemical potential $\mu_k - \mu_0$ on the quantities T, p, and c_n ($n = 1, \ldots, N$). However, it is more systematic to eliminate the quantity $\nabla(\mu_k - \mu_0)$ from (13.9) using (7.15). In this case, we may make use of the fact that the partial pressures p_k depend on T, $1/\rho_k$, and $\mathbf{E}^2$, without making assumptions concerning the character of the dependence of $\mu_k - \mu_0$ on the quantities T, p, c_n $(n = 1, \ldots, N)$.

Consider a model for the mixture where the following relations are satisfied (in this case, we will neglect effects connected with electrostriction):

$$p_k = p_k(T, 1/\rho_k, \mathbf{E}^2), \quad k = 0, \ldots, N, \tag{13.11}$$

$$\rho = \rho(T, p, c_1, \ldots, c_N), \quad \frac{\partial \varkappa_k}{\partial \rho_k} \equiv 0, \quad k = 0, \ldots, N. \tag{13.12}$$

Analogously to the derivation of (12.10) and (12.11), we obtain

$$A_k = A_k^{(0)} - \frac{1}{2}\varepsilon_* \varkappa_k \frac{1}{\rho}\mathbf{E}^2, \quad A_k^{(0)} = A(T, 1/\rho_k), \quad k = 0, \ldots, N, \tag{13.13}$$

$$p_k = p_k^{(0)} + \frac{1}{2}\varepsilon_* \varkappa_k c_k \mathbf{E}^2, \quad p_k^{(0)} = -\frac{\partial A_k^{(0)}}{\partial 1/\rho_k}, \quad p_k^{(0)} = p_k^{(0)}(T, 1/\rho_k), \tag{13.14}$$

$$\nabla p_k = \nabla p_k^{(0)} + \frac{1}{2}\varepsilon_* \nabla(c_k \varkappa_k)\mathbf{E}^2 + \varepsilon_* \varkappa_k c_k(\nabla \mathbf{E}) \cdot \mathbf{E}, \tag{13.15}$$

where $p_k^{(0)}$ is the partial pressure of the kth component in the absence of an electromagnetic field. Substituting (13.15) and (7.15), taking into account (12.3) and (13.4) and the relationships

$$\nabla \rho_k = \rho \nabla c_k + c_k \nabla \rho, \quad \nabla \rho = \frac{\partial \rho}{\partial T}\nabla T + \frac{\partial \rho}{\partial p}\nabla p$$

$$+ \sum_{n=1}^{N}\frac{\partial \rho}{\partial c_n}\nabla c_n, \quad \nabla c_0 = -\sum_{n=1}^{N}\nabla c_n \tag{13.16}$$

we obtain

$$\nabla\frac{\mu_k - \mu_0}{T} = \left\{-\frac{s_k - s_0}{T} + \frac{\varepsilon_*}{2\rho T}\mathbf{E}^2\frac{\partial(\varkappa_k - \varkappa_0)}{\partial T}\right.$$

$$+ \frac{1}{T}\left(\frac{1}{\rho_k}\frac{\partial p_k^{(0)}}{\partial T} - \frac{1}{\rho_0}\frac{\partial p_0^{(0)}}{\partial T}\right) + \frac{1}{T\rho}\frac{\partial \rho}{\partial T}\left(\frac{\partial p_k^{(0)}}{\partial \rho_k} - \frac{\partial p_0^{(0)}}{\partial \rho_0}\right)$$

$$\left.- \frac{\mu_k - \mu_0}{T^2}\right\} \cdot \nabla T + \frac{1}{\rho T}\left(\frac{\partial p_k^{(0)}}{\partial \rho_k} - \frac{\partial p_0^{(0)}}{\partial \rho_0}\right)\frac{\partial \rho}{\partial p}\nabla p +$$

$$+ \frac{1}{T} \left\{ \frac{1}{\rho} \frac{\partial p_k^{(0)}}{\partial \rho_k} \frac{\partial \rho}{\partial c_k} + \frac{1}{c_k} \left(\frac{\partial p_k^{(0)}}{\partial \rho_k} + \frac{1}{2} \frac{\varepsilon_* \varkappa_k}{\rho} \mathbf{E}^2 \right) - \frac{1}{\rho} \frac{\partial p_0^{(0)}}{\partial \rho_0} \frac{\partial \rho}{\partial c_k} \right.$$

$$\left. + \frac{1}{c_0} \left(\frac{\partial p_0^{(0)}}{\partial \rho_0} + \frac{1}{2} \frac{\varepsilon_* \varkappa_0}{\rho} \mathbf{E}^2 \right) \right\} \nabla c_k + \sum_{\substack{k,n \geqslant 1 \\ k \neq n}}^{N} \frac{1}{T} \left\{ \frac{1}{\rho} \left(\frac{\partial p_k^{(0)}}{\partial \rho_k} - \frac{\partial p_0^{(0)}}{\partial \rho_0} \right) \right.$$

$$\times \frac{\partial \rho}{\partial c_n} + \frac{1}{c_0} \left(\frac{\partial p_0^{(0)}}{\partial \rho_0} + \frac{1}{2} \frac{\varepsilon_* \varkappa_0}{\rho} \mathbf{E}^2 \right) \right\} \nabla c_n, \quad k = 1, \ldots, N. \quad (13.17)$$

Obviously, the kinetic coefficients L_{kk} in (13.9) should go to zero when $c_k = 0$ ($k = 1, \ldots, N$). In the opposite case, where $c_k \neq 0$ (i.e., in the absence of the kth component), the mass fluxes $\mathbf{i}_k$ may be different from zero. By virtue of this fact, when we substitute (13.17) into (13.9), we may neglect the last term in expression (13.17). Note that it is necessary to do this because we have excluded effects connected with cross-diffusion between components. Set

$$L_{kk} = L_{kk}^* \rho_k = L_{kk}^* \rho c_k \qquad (13.18)$$

and introduce the symbols

$$D_k = L_{kk}^* \frac{1}{T} \left\{ c_k \frac{1}{\rho} \left(\frac{\partial p_k^{(0)}}{\partial \rho_k} - \frac{\partial p_0^{(0)}}{\partial \rho_0} \right) \frac{\partial \rho}{\partial c_k} \right.$$

$$\left. + \frac{c_k}{c_0} \left(\frac{\partial p_0^{(0)}}{\partial \rho_0} + \frac{1}{2} \frac{\varepsilon_* \varkappa_0}{\rho} \mathbf{E}^2 \right) + \frac{\partial p_k^{(0)}}{\partial \rho_k} + \frac{1}{2} \varepsilon_* \frac{\varkappa_k}{\rho} \mathbf{E}^2 \right\} \qquad (13.19)$$

$$D_k^T = L_{kk}^* \frac{1}{T} \left\{ - \frac{h_k - h_0}{T} + \frac{1}{2} \varepsilon_* \mathbf{E}^2 \frac{\partial (\varkappa_k - \varkappa_0)}{\rho \partial T} \right.$$

$$\left. + \left(\frac{1}{\rho_k} \frac{\partial p_k^{(0)}}{\partial T} - \frac{1}{\rho_0} \frac{\partial p_0^{(0)}}{\partial T} \right) + \frac{1}{\rho} \left(\frac{\partial p_k^{(0)}}{\partial \rho_k} - \frac{\partial p_0^{(0)}}{\partial \rho_0} \right) \frac{\partial \rho}{\partial T} \right\}, \quad (13.20)$$

$$D_k^p = L_{kk}^* \frac{1}{T\rho} \left(\frac{\partial p_k^{(0)}}{\partial \rho_k} - \frac{\partial p_0^{(0)}}{\partial \rho_0} \right) \frac{\partial \rho}{\partial p}, \qquad (13.21)$$

$$\gamma_k = L_{kk}^* \frac{e_k - e_0}{T}, \qquad (13.22)$$

$$\chi = + \frac{L_{qq}}{T^2} . \tag{13.23}$$

Here, $L_{kk}*/T$ is the mobility of the kth component relative to the solvent (the 0th component); D_k is the diffusion coefficient; D_k^T is the thermal diffusion coefficient; D_k^p is the barodiffusion coefficient; γ_k is the electric mobility of the kth component relative to the solvent; χ is the thermal conductivity coefficient of the mixture. Finally, the defining relations have the form

$$\mathbf{i}_k = -\rho D_k \nabla c_k + \rho \gamma_k c_k \mathbf{E} - \rho D_k^p c_k \nabla p - \rho D_k^T c_k \nabla T + \rho \frac{L_{kk}^*}{T} c_k (\mathbf{F}_k - \mathbf{F}_0), \tag{13.24}$$

$$\mathbf{q} = -\chi \nabla T. \tag{13.25}$$

In concluding this section, let us show how (11.10) may be transformed to determine the temperature if relation (13.12) is satisfied. First, transform the last term on the right-hand side of (11.10), taking into account (11.4) and (13.12):

$$-\rho \sum_{k=0}^{N} c_k \left\{ T \frac{\partial p_k}{\partial T} \frac{d}{dt} \left(\frac{1}{\rho c_k} \right) + T \frac{\partial p_k}{\partial T} \cdot \frac{d\mathbf{E}}{dt} \right\} = -\rho T \frac{\partial p^e}{\partial T} \left\{ \frac{\partial}{\partial T} \left(\frac{1}{\rho} \right) \frac{dT}{dt} \right\}$$

$$+ \left\{ -\rho T \frac{\partial p^e}{\partial T} \frac{\partial}{\partial p} \left(\frac{1}{\rho} \right) \frac{dp}{dt} - \rho T \frac{\partial p^e}{\partial T} \sum_{k=1}^{N} \frac{\partial}{\partial c_k} \left(\frac{1}{\rho} \right) \frac{dc_k}{dt} \right.$$

$$+ \rho \sum_{k=0}^{N} \frac{1}{\rho_k} T \frac{\partial p_k}{\partial T} \frac{dc_k}{dt} - \rho T \frac{\partial p}{\partial T} \cdot \frac{d\mathbf{E}}{dt} \right\} \tag{13.26}$$

It is well known that (see, for example, [29])

$$c_p - c_v = T \frac{\partial p^e}{\partial T} \frac{\partial}{\partial T} (1/\rho), \tag{13.27}$$

where c_p is the specific heat of the mixture at constant pressure. Neglecting all the terms except the first one on the right-hand side of expression (13.26) yields

$$\rho c_p \frac{dT}{dt} + \operatorname{div} \mathbf{q} = \rho r + \sum_{k=0}^{N} e_k \mathbf{i}_k \cdot \mathbf{E} + \mathbf{T}^v : \nabla \mathbf{v} + \sum_{k=0}^{N} \mathbf{F}_k \cdot \mathbf{i}_k$$

$$- \frac{1}{2} \sum_{k=0}^{N} \sigma_k \mathbf{w}_k^2 - \rho \sum_{k=0}^{N} h_k \frac{dc_k}{dt} . \tag{13.28}$$

Usually the contribution to the heat source from the discarded terms is much smaller than, for example, that from the Joule effect. Note that the last term on the right-hand side of (13.28) describes the heat of the chemical reactions.

14. Complete System of Equations for Describing Multicomponent Mixtures

The fundamental hypotheses forming the basis for the model used for a multicomponent mixture is as follows. For the mixture as a whole, the law of conservation of mass is satisfied [see (1.9)]. The overall momentum of the internal forces is balanced by the overall momentum arising upon motion of the mass sources [see (2.5)]. In the equations of motion, we neglect the inertial forces arising upon motion of the components relative to the mixture, and also the momentum transport upon motion of the mass source relative to the mixture [see (10.1)]. The law of conservation of energy is used in form (5.4) [or (5.6)]; the Clausius–Duhem inequality is assumed to be satisfied for each component individually [see (6.2)].

To describe the electromagnetic field and its interaction with matter, Maxwell's equations with sources are used, i.e., charges and currents [see (4.1)-(4.6), (4.9), (4.10), (4.12)]. We assume that the magnetic field is weak and slowly varying, and omit magnetization of the mixture [see Section 12 and (11.3)]. Neither nonlinear effects arising upon polarization [see (12.8)] nor the contribution from electrostriction to the mass fluxes of the components [see (10.12)] are included. The forces are due to the electromagnetic field on the components of the mixture as described by relation (4.32). In this case, we will use the form (12.15) for the forces acting on the mixture as a whole upon its polarization. The viscous stress tensor is specified for the mixture as a whole [see (13.1)]. In the expressions for the heat and mass fluxes, all possible crossover effects are neglected [see (13.24), (13.25), and Section 13].

In the equation for determining the temperature, (11.10), we neglect the heat sources connected with the work done by viscous stresses (the term $\mathbf{T}^v : \nabla \mathbf{v}$), and the heat sources due to the temperature dependence of the partial pressures p_k and the dielectric susceptibilities $\varkappa_k$ [see (13.14)]. Furthermore, we assume that the solvent (the 0th component) is electrically neutral, and the specific charges of the rest of the components are constant (possibly zero):

$$e_0 = 0, \quad e_k = \text{const}, \quad k = 1, \ldots, N. \tag{14.1}$$

We assume that the external forces acting on the mixture are gravitational forces:

$$\mathbf{F}_k = \mathbf{g}, \quad k = 0, \ldots, N, \tag{14.2}$$

where $\mathbf{g}$ is the acceleration of gravity.

As the parameters for describing the mixture we choose the following $N + 8$ quantities: the density of the mixture ρ, velocity of the mixture $\mathbf{v}$, concentrations of the components c_k ($k = 1, \ldots, N$), electric field intensity $\mathbf{E}$, and temperature T. The complete system of equations has the form

$$\frac{d\rho}{dt} + \rho \operatorname{div} \mathbf{v} = 0, \tag{14.3}$$

$$\rho \frac{\partial c_k}{dt} + \operatorname{div} \mathbf{i}_k = \sigma_k (c_1, \ldots, c_N), \quad k = 1, \ldots, N, \tag{14.4}$$

$$\mathbf{i}_k = -\rho D_k \nabla c_k + \rho c_k \gamma_k \mathbf{E} - \rho D_k^p c_k \nabla p - \rho D_k^T c_k \nabla T, \quad k = 1, \ldots, N, \tag{14.5}$$

$$\rho \frac{d\mathbf{v}}{dt} = -\nabla p + \operatorname{div} \mathbf{T}^v + \rho \mathbf{g} - \frac{1}{2} \varepsilon_* (\nabla \varepsilon) \mathbf{E}^2 + \frac{1}{2} \nabla \left(\rho \frac{\partial \varepsilon}{\partial \rho} \varepsilon_* \mathbf{E}^2 \right)$$
$$+ \rho \sum_{k=1}^N c_k e_k \mathbf{E}, \tag{14.6}$$

$$\mathbf{T}^{v,\alpha\beta} = \eta \left(\frac{\partial v^\alpha}{\partial x_\beta} + \frac{\partial v^\beta}{\partial x_\alpha} \right) - \left(\frac{2}{3} \eta - \eta_v \right) \frac{\partial v^\gamma}{\partial x_\gamma} \delta_{\alpha\beta}, \tag{14.7}$$

$$\varepsilon_* \frac{\partial}{\partial t} (\varepsilon \mathbf{E}) + \operatorname{rot} \sum_{k=0}^N \{ c_k \varkappa_k \varepsilon_* \mathbf{E} \wedge (\mathbf{v} + \mathbf{w}_k) \} + \sum_{k=1}^N e_k \mathbf{i}_k + \rho \sum_{k=1}^N c_k e_k \mathbf{v} = \mathbf{J}, \tag{14.8}$$

$$\varepsilon = \varepsilon_0 + \sum_{k=1}^N (\varepsilon_k - \varepsilon_0) c_k, \quad \mathbf{w}_k = \frac{\mathbf{i}_k}{\rho c_k}, \quad c_0 = 1 - \sum_{k=1}^N c_k, \quad \operatorname{div} \mathbf{J} = 0, \tag{14.9}$$

$$\varepsilon_* \operatorname{div} (\varepsilon \mathbf{E}) = \rho \sum_{k=1}^N c_k e_k, \quad \operatorname{rot} \mathbf{E} = 0, \tag{14.10}$$

$$\rho c_p \frac{dT}{dt} + \operatorname{div} \mathbf{q} = \rho r + \sum_{k=1}^N e_k \mathbf{i}_k \cdot \mathbf{E} - \frac{1}{2} \sum_{k=1}^N (\mathbf{w}_k^2 - \mathbf{w}_0^2)$$
$$- \rho \sum_{k=1}^N (h_k - h_0) (\sigma_k - \operatorname{div} \mathbf{i}_k), \tag{14.11}$$

$$\mathbf{q} = \varkappa \nabla T, \tag{14.12}$$

$$c_p = c_{v,0} + \sum_{k=1}^N c_k (c_{v,k} - c_{v,0}) + T \frac{\partial p^e}{\partial T} \frac{\partial}{\partial T} \left(\frac{1}{\rho} \right). \tag{14.13}$$

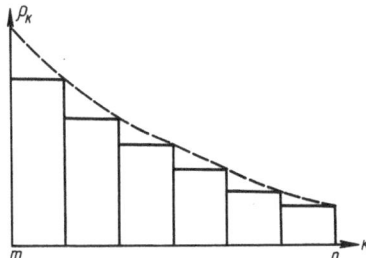

Fig. 2. Replacement of the discrete density
distribution by a continuous density distribution
(one-dimensional case).

The quantities $\mathbf{J}$, ε_k, ε_*, $\varkappa_k$, e_k, $D_k{}^p$, $D_k{}^T$, D_k, γ_k, h_k, η, η_v, χ, ρr,

$T \dfrac{\partial p^e}{\partial T} \dfrac{\partial}{\partial T} \left(\dfrac{1}{\rho} \right)$, $c_{v,k}$ and their dependence on T, $\mathbf{E}$, c_1, ..., c_N are assumed
to be known. The expressions for the mass sources σ_k ($k = 1, ..., N$) will
be written out for each concrete case (see below).

15.   Infinite-Component Mixtures

The mathematical description of the mixture is complicated as the
number of components and the number of chemical reactions increases. At the
same time, especially when we consider biopolymers and processes connected with them, the number of components is measured in the thousands,
and the chemical reactions are quite diverse and not always known. Such
complicated systems can hardly be solved by contemporary means; moreover, it is not clear what we would do with such a vast amount of information even if we could obtain it.

As has been indicated in the introductory part of this chapter, a
promising route for shortening the description is connected with introducing models for mixtures consisting of an infinite number of components;
and in the theory interest is focused not so much on the behavior of the individual components as on the behavior of large groups of these components which are jointly characterized by similar values for essential parameters. The generalization to a countable set of components occurs very directly. In order to obtain such a model, simply set $N = \infty$ in the preceding
treatment and replace the sums over all components by infinite series. For
example, for the density of the mixture ρ_n we have

$$\rho_n (x, \ t) = \sum_{k=0}^{\infty} \rho_k (x, \ t). \qquad (15.1)$$

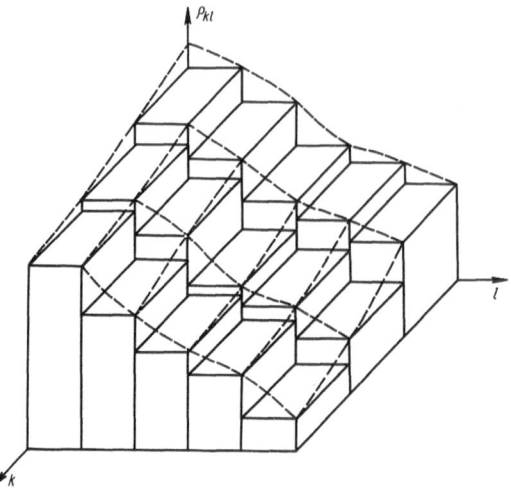

Fig. 3. Replacement of a discrete density distribution by a continuous density distribution (two-dimensional case).

Now consider the case when the quantity ρ_k weakly varies with a change in k. In this case, it is no longer expedient to consider the quantities ρ_k separately; focus instead on the contribution to the density (and other quantities) from components with numerical labels included within some limits, e.g., $m < k < n$. To a good approximation the sum $\sum_{k=m}^{n}$ may be replaced by the integral $\int_{m}^{n} \rho(k)dk$ (Fig. 2), assuming now that $\rho(k)$ is a function determined not only for integral k but in general for all real, positive k. Then, instead of (15.1), we will have

$$\rho_n(x,\ t) = \int_0^{\infty} \rho(k,\ x,\ t)\,dk. \tag{15.2}$$

A typical example of this is a linear polymer. In this case, k designates the number of monomer units in the polymer molecule or the length of the polymer chain. It is not difficult to represent cases when there are two essential parameters, where it is expedient to write series (15.1) in the form

$$\rho_n(x,\ t) = \sum_k \sum_e \rho_{ke}(x,\ t), \tag{15.3}$$

and ρ_{ke} varies weakly with a change in k or e and the sum (15.3) is replaced by a double integral (Fig. 3):

$$\rho_n(x, \ t) = \int\limits_0^\infty \int\limits_0^\infty \rho(k, \ e, \ x, \ t) \, dk dl. \qquad (15.4)$$

In complicated cases there may be many such parameters; they are called sorting parameters (of the particles). By this means, of course, we arrive at the following generalization, including the idea that the S-space is the space of the sorting parameters.

Let S be the space with (positive, countably additive) measure $\mu(S)$, called the S-space in the following. To each point $s \in S$ corresponds a component of the mixture; and if a μ-measurable set of components $E \subset S$ is isolated, then their contribution to the general density is

$$\int\limits_E \rho(s, \ x, \ t) \, d\mu(s) \qquad (15.5)$$

(the Lebesgue–Stieltjes integral). In this case, for the total density we have

$$\rho_n(x, \ t) = \int\limits_S \rho(s, \ x, \ t) \, d\mu(s). \qquad (15.6)$$

These equations are directly generalized to arbitrary mixtures which, generally speaking, are infinite-component; the parameter k (the label of the mixture component) is replaced in the partial equations by $s \in S$, and the sum over k is replaced by integrals over s relative to the measure $\mu(s)$. Because this transition is obvious, we will not write out these equations here. We note that, from a logical standpoint, the described transition is a postulate. We will return to the case of infinite-component mixtures where S is a finite set, and the measure μ for each single-point set takes on the value 1.

The theory of infinite-component mixtures is included in the theory of internal degrees of freedom developed within the framework of nonequilibrium thermodynamics (see [11, 38]), where in fact essentially continuous parameters such as the length of a deformable molecule, the angle of rotation of a molecule in a field, etc., are considered as internal parameters. In a number of papers (see [2, 8], where further references may be found), the idea of making a discrete parameter (the length of the polymeric chain) continuous is used to describe polymerization processes.

The approach described for constructing a theory of infinite-component mixtures opens up new analytical possibilities. Chemical reactions are treated as motion in S-space; and in contrast to motion in physical space, here both continuous motion (for example, reactions in which a molecule of a linear polymer after a "short" time successively adds or loses a few units) as well as discrete jumps (adhesion or splitting of the polymer molecules) are possible. Study of the general laws of motion in S-space may possibly some day allow us to determine the fundamental general principles of the now immense set of reactions that occur in parallel in biological media. It would be of interest to investigate a number of ideas connected with the S-space, especially generalized mass flux, energy flux, etc., and also the corresponding analogs for dissipative structures.

Chapter II

CHEMICAL SUBSYSTEMS

The system of equations (I.14.3)-(I.14.13) is still too complicated for most purposes. The most fundamental approach to simplification is based on separating all the processes according to their rates. In fact, in most cases the characteristic diffusion time is significantly greater than the characteristic time connected with transport processes due to an electric field. The latter is, in turn, much longer than the characteristic time for occurrence of chemical reactions in solutions.

A natural route for simplifying the equations describing such "mixed-rate" processes is to separate the so-called slow variables, which vary weakly during the characteristic times for occurrence of all the rest of the processes. The asymptotic relationships connecting the "slow" variables with the rest of the "fast" variables in the simplest case prove to be algebraic and may be obtained by considering the corresponding steady-state equations. The existence of relationships between the instantaneous values of the "fast" and "slow" variables allows us to substantially simplify the description; the "fast" variables are eliminated from the system using algebraic relationships so as to reduce the problem to investigating the evolution of "slow" variables alone.

We will designate as "mixtures with stabilized local chemical equilibrium" those mixtures in which the characteristic time for occurrence of chemical reactions is much shorter than the characteristic times for all other processes [16]. In the investigation of the processes of heat, mass, and charge transport in such mixtures, we may assume that at each point of the mixture at any given instant of time chemical equilibrium exists between its components. Consequently, a number of algebraic relationships are satisfied, viz. equilibrium conditions for the concentrations containing the rate constants for the chemical reactions. Generally speaking, the latter depend on temperature, electric field, and some concentrations, which is one reason why the equilibrium proves to be only local. We note that different kinds of slow variables are possible. First of all, we have integrals of the chemical kinetics equations; secondly, we have the equilibrium concentrations for irreversible chemical reactions.

As a result of the reduction, we arrive at a system of equations which may be interpreted as the mathematical model for a new mixture whose components are sets of the original components with concentrations interconnected by algebraic relationships. These sets are called chemical subsystems in the following development. We emphasize that the idea of a chemical subsystem is not a purely mathematical abstraction. In the experiment, chemical subsystems should appear as individual substances with specific physical properties; a substance dissociating to ions in solution provides the most obvious and widespread example of a chemical subsystem. However, more complicated kinds of chemical subsystems exist.

An important characteristic of the new mixture is the absence of chemical reactions. This is not surprising since they have already been taken into account. We further note that the same component of the original mixture may be distributed over several chemical subsystems.

Of course, the hypothesis concerning local chemical equilibrium has a limited region of application. We may encounter situations where the rate of several chemical reactions is comparable with the rate of the transport processes. In this case, the model is correspondingly complicated, and the differential balance equations for several components are not reduced to algebraic equations.

In this chapter, we indicate the integrals of the chemical kinetics equations which are chosen as the slow variable for reversible isothermal reactions. We show that in the specific case of aqueous solutions it is feasible and expedient to choose the acidity of the solution as one of the slow variables. Finally, dimensionless variables are introduced and the conditions for applicability of the model obtained are considered.

1. Integrals for the Chemical Kinetics Equations Describing Equilibrium Chemical Reactions

Let us consider a chemically reactive $(N + 1)$-component, dilute aqueous solution. Designate the concentration (mass fraction) of the solvent (water) as $c_0 \approx 1$, and the concentrations of the solutes as $c_k \ll c_0$ ($k =$

1, ..., N). Assume that there are no processes other than chemical reactions occurring in the mixture. We write the chemical kinetics equations as

$$\rho \frac{\partial c_k}{\partial t} = \sigma_k \quad (k = 0, \ldots, N), \quad \sum_{k=0}^{N} \sigma_k = 0, \quad \sum_{k=0}^{N} c_k = 1, \qquad (1.1)$$

where σ_k is the mass source density for the kth component arising as the result of chemical reactions. Next, designate ρ_* as the average density of the solution, and let us introduce the molar concentrations of the mixture components:

$$a_k = \frac{c_k \rho_*}{m_k N_A}, \quad k = 0, \ldots, N, \qquad (1.2)$$

where m_k is the mass of a single structural unit (for example, a molecule or an ion) of the kth component; N_A is Avogadro's number. Let $R < N$ reversible chemical reactions (dissociation reactions, bimolecular reactions, etc.) of the following type occur in the mixture:

$$\mu_l^r a_l + \mu_p^r a_p \underset{k_r^-}{\overset{k_r^+}{\rightleftharpoons}} \nu^m a_m + \nu_n^r a_n, \quad r = 1, , \ldots, R < N,$$

$$(1.3)$$

$$l, \ p, \ m, \ n = 0, 1, \ldots, N,$$

where μ_l^r, μ_p^r, ν_m^r, and ν_n^r are stoichiometric coefficients for the rth reaction; k_r^+ and k_r^- are the rate constants, respectively, for the forward and reverse rth reactions.

We assign the index R to the dissociation reaction (autoprotolysis) of the solvent if it occurs in the mixture and designate the molar concentrations of the hydronium ion H_3^+O (in some cases, protons H^+, depending on the method of writing the reaction) and the hydroxyl ion OH^- by a_1 and a_2. Thus

$$\mu_1^R a_1 + \mu_2^R a_2 \underset{k_R^-}{\overset{k_R^+}{\rightleftharpoons}} \nu_0^R a_0, \quad \mu_1^R \equiv \mu_2^R \equiv 1. \qquad (1.4)$$

Then, since the solution is dilute, the activity of the kth component is equal to its molar concentration in all the chemical kinetics equations. The source of the rth reaction (the number of moles converted in the reaction per unit time per unit volume), which is equal to the difference between the rates of the reverse and forward reactions (see, for example, [10, 56]), is introduced as

$$\sigma^{(r)} = - k_r^+ a_l^{\mu_l^r} a_p^{\mu_p^r} + k_r^- a_m^{v_m^r} a_n^{v_n^r}, \quad r = 1, \ldots, R. \tag{1.5}$$

Then the mass source density in (1.1) may be represented as

$$\sigma_k = m_k N_A \sum_{r=1}^{R} (v_k^r - \mu_k^r) \sigma^{(r)}, \quad k = 0, \ldots, N. \tag{1.6}$$

We use I_k to designate the set of indices, not including the index 0, for which the following relationship is satisfied:

$$\sum_{\in I_k} \alpha_s^k \frac{\rho_* \sigma_s}{m_s N_A} \equiv \sum_{r=1}^{R} \rho_* \sigma^{(r)} \sum_{s \in I_k} \alpha_s^k (v_s^r - \mu_s^r) = 0, \quad k = 1, 2, \ldots, N_0, \tag{1.7}$$

where α_s^k denotes constant coefficients. We call the set of concentrations

$$A_k = \{a_s : s \in I_k\}, \quad k = 1, 2, \ldots, N_0, \tag{1.8}$$

the chemical subsystem. Condition (1.7) means that the number of particles (the number of moles) in the chemical subsystem is conserved as the chemical reactions occur.

Multiplying (1.1) by $\alpha_s^k \rho_*/m_k N_A$ and summing over $s \in I_k$, taking into account (1.2) and (1.7), we obtain

$$\sum_{s \in I_k} \frac{\partial (\alpha_s^k a_s)}{\partial t} = 0. \tag{1.9}$$

Introduce the molar concentration of the chemical subsystem as

$$\xi_k = \sum_{s \in I_k} \alpha_s^k a_s, \quad k = 1, 2, \ldots, N_0. \tag{1.10}$$

From (1.9), it follows that

$$\frac{\partial \xi_k}{\partial t} = 0, \quad k = 1, 2, \ldots, N_0. \tag{1.11}$$

Thus, the quantities ξ_k ($k = 1, 2, \ldots, N_0$) are integrals for the system of chemical kinetics equations (1.1).

It can be shown that system (1.1) has at least one integral, if the solution under consideration contains charged particles. Let z_s ($s = 0, 1, ...,$ N), the charge of the sth component of the mixture, be the number of electrons with allowance for the sign of the charge per structural unit (ion) of the sth component. Designate the set of indices for the charged components as

$$I_0 = \{s : z_s \neq 0\}. \tag{1.12}$$

The specific charge of the sth component is

$$e_s = \frac{ez_s}{m_s}, \quad s \in I_0. \tag{1.13}$$

From the law of conservation of charge during the chemical reactions,

$$\sum_{s \in I_0} e_s \sigma_s = 0, \tag{1.14}$$

it follows that

$$\frac{\rho_*}{N_A} \sum_{s \in I_0} e_s \sigma_s \equiv \sum_{s \in I_0} \frac{ez_s \rho_*}{m_s N_A} \sigma_s \equiv e \sum_{s \in I_0} z_s \frac{\rho_* \sigma_s}{m_s N_A} = 0. \tag{1.15}$$

Comparing the expression obtained with (1.7), and introducing, analogous to (1.10), the quantity ξ_0, which we will call the molar charge of the solution, we obtain

$$\xi_0 = \sum_{s \in I_0} z_s \alpha_s^0, \quad \alpha_s^0 \equiv z_s, \tag{1.16}$$

$$\frac{\partial \xi_0}{\partial t} = 0. \tag{1.17}$$

Thus, the quantity ξ_0 is the integral for system (1.1).

Next, consider the case where the solution is found to be in chemical equilibrium, i.e., the sources of all the rth reactions go to zero:

$$\sigma^{(r)} = 0, \quad r = 1, \ldots, R. \tag{1.18}$$

Using (1.3), for the equilibrium concentrations we obtain the relationships (recall that the solution is dilute and molar concentrations have replaced the activities)

$$\frac{a_m^{\nu_m^r} a_n^{\nu_n^r}}{a_l^{\mu_l^r} a_p^{\mu_p^r}} = \frac{k_r^+}{k_r^-} \equiv \tilde{K}_r, \quad r = 1, \ldots, R, \tag{1.19}$$

where $\tilde{K}_r$ is the concentration equilibrium constant of the rth reaction. In the general case, $\tilde{K}_r$ depends on the ionic strength of the solution (see, for example, [23]).

If one of the indices m, n, l, p in (1.5) is equal to zero, for example $l = 0$, then we will write the expression (1.19) in the form

$$\frac{a_m^{v_m'} a_n^{v_n'}}{a_p^{\mu_p'}} = K_r \equiv \tilde{K}_r \cdot a_0^{v_0'}. \tag{1.20}$$

The quantity K_r is called the reduced equilibrium constant for the rth reaction. If in (1.5) we have $m, n, l, p \neq 0$, then $K_r \equiv \tilde{K}_r$.

In the special case of the autoprotolysis reaction of the solvent, cf. (1.4), we have

$$a_1 \cdot a_2 = \tilde{K}_R a_0^{v_0^R} \equiv K_w^2, \tag{1.21}$$

where K_w^2 is the ion product of water.

Let the system (1.1) have $N - R$ linearly independent integrals $\xi_0, \xi_1,$ $\ldots, \xi_{N-R-1}$ ($N_0 \equiv N - R - 1$). Introduce the function

$$\Phi_k (a_1, \ldots a_N; \xi_0, \ldots, \xi_{N-R-1}) \equiv \xi_{k-1} - \sum_{s \in I_{k-1}} \alpha_s^{k-1} a_s, \tag{1.22}$$
$$k = 1, \ldots, N - R,$$

$$\Phi_k (a_1, \ldots, a_N; \xi_0, \ldots, \xi_{N-R-1}) \equiv \sigma^{(k+R-N)} (a_1, \ldots, a_N), \tag{1.23}$$
$$k = N - R + 1, \ldots, N.$$

Consider the system of equations

$$\Phi_k (a_1, \ldots, a_N; \xi_0, \ldots, \xi_{N-R-1}) = 0, \quad k = 1, \ldots, N. \tag{1.24}$$

Suppose the Jacobian of the system is different from zero:

$$J \equiv \frac{D (\Phi_1, \ldots, \Phi_N)}{D (a_1, \ldots, a_N)} \neq 0. \tag{1.25}$$

Then, according to the implicit function theorem, the system of equations (1.24) defines the $a_1, \ldots, a_N$ as single-valued functions of the arguments $\xi_0, \ldots, \xi_{N-R-1}$:

$$a_k = \varphi_k (\xi_0, \ldots, \xi_{N-R-1}); \; k = 1, \ldots, N. \tag{1.26}$$

In this case, the functions φ_k are continuous and have partial derivatives with respect to all the arguments. Thus, the molar concentrations of the solutes may be expressed in terms of the molar concentrations of the $N - R - 1$ chemical subsystems and the molar charge of the solution.

For the following development, it is convenient to slightly change the form of writing relationships (1.26) by introducing the equilibrium degrees of dissociation $\beta_k{}^s$, the relative concentrations of the sth component of the solution in the kth chemical subsystem:

$$\beta_k^s = \frac{\alpha_s^k \varphi_s\,(\xi_0,\ \ldots,\ \xi_{N-R-1})}{\xi_k}, \quad k = 0,\ \ldots,\ N - R - 1,\ s \in I_k. \quad (1.27)$$

Then (1.26) takes on the form

$$a_s = \frac{\beta_k^s}{\alpha_s^k}\,\xi_k,\quad \beta_k^s = \beta_k^s\,(\xi_0,\ \ldots,\ \xi_{N-R-1})\ s \in I_k,\ k = 0,\ \ldots,\ N - R - 1.$$
$$(1.28)$$

It is not difficult to convince ourselves, using (1.10) and (1.16), that

$$\sum_{s \in I_k} \beta_k^s = 1 \quad k = 0,\ \ldots,\ N - R - 1. \quad (1.29)$$

It is also convenient to introduce the following sets of indices which will be used later on:

$$I_{01} = I_0 \cap I_1,\quad I_{02} = (I_0 \setminus I_{01}) \cap I_2,\quad I_{03} = ((I_0 \setminus I_{01}) \setminus I_{02}) \cap I_3,\ \ldots$$
$$\ldots,\ I_{0,N-R} = ((I_0 \setminus I_{01}) \setminus I_{02} \setminus \ \ldots\ ) \cap I_{N-R-1},$$
$$I_{00} = I_0 \setminus (I_{01} \cup I_{02} \cup\ \ldots\ \cup I_{0,N-R-1}). \quad (1.30)$$

Using these symbols, we may, for example, write the molar charge distributions as

$$\xi_0 \equiv \sum_{s \in I_0} z_s a_s = \sum_{k=0}^{N-R-1} \left(\sum_{s \in I_{0k}} z_s\,\frac{\beta_k^s}{\alpha_s^k} \right) \xi_k \equiv \varphi_0\,(\xi_0,\ \xi_1,\ \ldots,\ \xi_{N-R-1}). \quad (1.31)$$

2. Integrals of the Chemical Kinetics Equations for "Slow" Variables for the Simplified System of Equations

Let us show that, under certain assumptions, the integrals of the chemical kinetics equations are the most natural "slow" variables, slowly varying as all the chemical reactions take place in the mixture. When we introduce such variables, we can divide the mixture into chemical subsystems which, from the standpoint of diffusion transport and similar processes, behave as an integrated whole. The multicomponent mixture of chemical subsystems obtained in this case may contain considerably fewer components than the original mixture. The processes occurring in the mixture of chemical subsystems are described in the general case by an open system of equations. This system can be closed if we assume local chemical equilibrium, i.e., the characteristic time for all the chemical reactions to occur in the original mixture is much less than the characteristic times for all the rest of the processes.

Let the $(N + 1)$-component mixture be a dilute aqueous solution, so that

$$c_0 \approx 1, \quad c_k \ll c_0, \quad k = 1, \ldots, N. \tag{2.1}$$

In the mixture, $R < N$ chemical reactions of type (1.3) occur. The mass source density for the kth components, σ_k, and the reaction sources, $\sigma^{(r)}$ $(r = 1, \ldots, R)$, have forms given by (1.5) and (1.6). We assume that the system of chemical kinetics equations (1.1) has $N - R$ independent integrals, i.e., conditions (1.7) are satisfied. Thus, the mixture is decomposed into chemical subsystems A_k $(k = 1, \ldots, N - R - 1)$ with molar concentrations ξ_k $(k = 1, \ldots, N - R - 1)$. The mixture is found to be in local chemical equilibrium; i.e., the sources $\sigma^{(r)}$ $(r = 1, \ldots, R)$ and consequently also σ_k $(k = 1, \ldots, N)$ are equal to zero.

The following assumptions are made for convenience and allow us to neglect several terms in (I.14.3)-(I.14.13). The only information we have about effects described by these terms is that, as a rule, they are not important in the study of processes occurring in mixtures. We will assume that the heat capacity of the mixture is equal to the heat capacity of the solvent; i.e., in (1.14) we neglect terms of the type $\sum_{k=1}^{N} c_k (c_{v,k} - c_{v,0}) +$

$T \dfrac{\partial p^e}{\partial T} \dfrac{\partial}{\partial T} \left(\dfrac{1}{\rho} \right)$ compared with $c_{v,0}$. In (I.14.6), we neglect the term $\nabla/2 \times$

$\left(\rho \dfrac{\partial \varepsilon}{\partial \rho} \varepsilon_* E^2 \right)$ by assuming that the dielectric constant of the solution does not

depend on density. We omit the bulk viscosity η_v in expression (I.14.7) and will take the solvent to be electrically neutral ($e_0 \equiv 0$).

Before we write the system of equations for describing a multicomponent mixture of chemical subsystems, we emphasize that such a description holds independent of the assumption of local chemical equilibrium due to Eqs. (I.14.3)-(I.14.13). The difference between this system and system (I.14.3)-(I.14.13) is the fact that the former is not closed. Recall that, in Section 1 of Chapter I, the mass sources are assumed to be known. The hypothesis of local chemical equilibrium allows us to close the system of equations for describing the mixture of chemical subsystems. The listed characteristics also pertain to the equation obtained from (I.14.11), in which, in connection with the hypothesis of local chemical equilibrium, we will discard terms of the type $-\frac{1}{2}\sum_{k=1}^{N} \sigma_k(w_k{}^2 - w_0{}^2) - \rho \sum_{k=1}^{N} (h_k - h_0)\sigma_k$.

Now introduce the degrees of dissociation (not necessarily equilibrium values) connecting the molar concentrations of the components of the solution with the molar concentrations of the chemical subsystems [compare with (1.28)]:

$$a_s = \frac{\beta_k^s}{\alpha_s^k} \xi_k, \quad s \in I_k, \quad k = 0, \ldots, N - R - 1,$$

$$\beta_k^s = \beta_k^s (\xi_0, \ldots, \xi_{N-R-1}; \; a_1, \ldots, a_N). \tag{2.2}$$

Choose the integrals of the chemical kinetics equations $\xi_0, \ldots, \xi_{N-R-1}$ from formulas (1.10) and (1.16) as the "slow" variables and, using (1.2) and (2.2), obtain

$$c_k = \frac{m_k N_A}{\rho_*} \beta_e^k \frac{1}{\alpha_k^l} \xi_e, \quad k \in I_l, \quad e = 1, \ldots, N - R - 1. \tag{2.3}$$

The molar flux of the kth chemical subsystem is

$$\mathbf{i}_{\xi_k} = \sum_{s \in I_k} \frac{\alpha_s^k}{m_s N_A} \mathbf{i}_s, \quad k = 1, \ldots, N - R - 1. \tag{2.4}$$

Next, introduce the diffusion, thermal diffusion, barodiffusion, and mobility coefficients for the kth chemical subsystem:

$$D_{\xi_k} \equiv \sum_{s \in I_k} D_s \beta_k^s, \quad D_{\xi_k}^T \equiv \sum_{s \in I_k} D_s^T \beta_k^s, \quad D_{\xi_k}^p \equiv \sum_{s \in I_k} D_s^p \beta_k^s,$$

$$\gamma_{\xi_k} \equiv \sum_{s \in I_k} \gamma_s \beta_k^s, \quad k = 1, \ldots, N - R - 1. \tag{2.5}$$

Then the molar flux of the kth chemical subsystem has the form

$$\mathbf{i}_{\xi_k} = \frac{\rho}{\rho_*} \{ -\nabla (D_{\xi_k} \xi_k) + \gamma_{\xi_k} \xi_k \mathbf{E} - D_{\xi_k}^T \xi_k \nabla T - D_{\xi_k}^p \xi_k \nabla p \} \tag{2.6}$$

$$k = 1, \ldots, N - R - 1.$$

Multiplying (I.14.4) by $(\rho_*/m_s N_A)\alpha_s^k$ and summing over all $s \in I_k$ ($k = 1, \ldots, N - R - 1$), using (1.2), (1.7), (1.10), and (2.4), we derive

$$\frac{\rho}{\rho_*} \frac{d\xi_k}{dt} + \operatorname{div} \mathbf{i}_{\xi_k} = 0, \quad k = 1, \ldots, N - R - 1. \tag{2.7}$$

The equation obtained is the law of conservation of number of particles (moles) of the chemical subsystem. Thus, from the standpoint of describing the transport processes, the original mixture has been divided into a mixture of $(N - R - 1)$ chemical subsystems of concentration ξ_k ($k = 1, \ldots, N - R - 1$).

For the chemical subsystems, let us introduce coefficients for the molar diffusion conductivity, barodiffusion, thermal diffusion, and molar conductivity:

$$\sigma^D_{\xi_k} \equiv \sum_{s \in I_{0k}} z_s D_s \frac{\beta^s_k}{\alpha^k_s}, \quad \sigma^p_{\xi_k} \equiv \sum_{s \in I_{0k}} z_s D^p_s \frac{\beta^s_k}{\alpha^k_s}, \quad \sigma^T_{\xi_k} \equiv \sum_{s \in I_{0k}} z_s D^T_s \frac{\beta^s_k}{\alpha^k_s},$$

$$\sigma_{\xi_k} \equiv \sum_{s \in I_{0k}} z_s \gamma_s \frac{\beta^s_k}{\alpha^k_s}, \quad k = 0, \ldots, N - R - 1. \tag{2.8}$$

Then the expression for the electric current density in solution has the form

$$\mathbf{j} = \frac{\rho}{\rho_*} e N_A \sum_{k=0}^{N-R-1} \{\nabla (\sigma^D_{\xi_k} \xi_k) + \sigma_{\xi_k} \xi_k \mathbf{E} - \sigma^T_{\xi_k} \xi_k \nabla T - \sigma^p_{\xi_k} \xi_k \nabla p\}. \tag{2.9}$$

Differentiate Eq. (I.14.10) with respect to time and take the divergence of (I.14.8). Comparing the expressions obtained, and taking into account (1.2), (1.16), (2.2), and (2.3), we derive

$$\frac{\rho}{\rho_*} \frac{d\xi_0}{dt} + \text{div } \mathbf{i}_{\xi_0} = 0, \quad \mathbf{i}_{\xi_0} \equiv \frac{1}{e N_A} \cdot \mathbf{j}. \tag{2.10}$$

This expression is the law of conservation of molar charge for the solution. Formally, the quantity ξ_0 may be considered as the concentration of one more chemical subsystem. Equation (I.14.10) now has the form

$$\varepsilon_* \text{ div } (\varepsilon \mathbf{E}) = \frac{\rho}{\rho_*} e N_A \xi_0, \quad \text{rot } \mathbf{E} = 0. \tag{2.11}$$

We also write Eqs. (I.14.6), (I.14.7), and (I.14.11), which in practice vary only slightly,

$$\rho \frac{d\mathbf{v}}{dt} = -\nabla p + \text{div } \mathbf{T}^v + \rho \mathbf{g} - \frac{1}{2} \varepsilon_* (\nabla \varepsilon) \mathbf{E}^2 + \frac{\rho}{\rho_*} e N_A \xi_0 \mathbf{E}, \tag{2.12}$$

$$\rho c_p \frac{dT}{dt} + \text{div } \mathbf{q} = \rho r + \mathbf{j} \cdot \mathbf{E}$$

$$+ \rho \sum_{s=1} (h_s - h_0) \text{ div } \mathbf{i}_s (\xi_0, \ldots, \xi_{N-R}; a_1 \ldots, a_N), \tag{2.13}$$

$$\mathbf{q} = -\chi \nabla T; \quad T^{v, \alpha\beta} = \eta \left(\frac{\partial v^\alpha}{\partial x^\beta} + \frac{\partial v^\beta}{\partial x^\alpha} \right) - \frac{2}{3} \eta \delta_{\alpha\beta} \frac{\partial v^\gamma}{\partial x^\gamma}. \tag{2.14}$$

The system of equations (I.14.3), (2.6), (2.7), (2.9)-(2.14) describes the behavior of the $(N - R)$-component mixture of chemical subsystems and this system is open. For closure we require that the following equalities be satisfied:

$$\sigma^{(r)}(a_1, \ldots, a_N) = 0, \quad r = 1, \ldots, R. \tag{2.15}$$

Next we determine the degrees of dissociation $\beta_k{}^s$ in formulas (2.2), (2.3), (2.5), (2.8) from system (1.24) [compare (1.26) and (1.27)]

$$\beta_k^s = \frac{\alpha_s^k \varphi_s (\xi_0, \ldots, \xi_{N-R-1})}{\xi_k}, \quad k = 0, \ldots, N - R - 1, \quad s \in I_k,$$

$$\beta_k^s = \beta_k^s (\xi_0, \ldots, \xi_{N-R-1}). \tag{2.16}$$

In this case, the quantity ε takes on the form

$$\varepsilon = \varepsilon_0 + \sum_{s=1} (\varepsilon_s - \varepsilon_0) \frac{m_s N_A}{\rho_*} \varphi_s (\xi_0, \ldots, \xi_{N-R-1}). \tag{2.17}$$

Assuming that the dependence of the quantities ρ, c_p, χ, η, ρr, ε_s, h_s ($s = 0$, ..., N) on T, ξ_0, ..., ξ_{N-R-1} is known, we obtain a completely closed system of equations for describing the processes occurring in multicomponent mixtures.

3. Transition to Dimensionless Variables

Let us write the system of equations in the case of local chemical equilibrium in dimensionless variables. For this, we make the replacements

$$\left.\begin{array}{l}
\rho \to \rho_* \rho, \quad x \to Lx, \quad t \to \mathcal{T}t, \quad E \to \mathcal{E}E, \quad T \to T_* T, \quad D_s \to D_* D_s, \\[2mm]
\xi_s \to K_* \xi_s, \quad a_s \to K_* a_s, \quad D_s^T \to \frac{D_*}{T_*} \cdot D_s^T, \quad D_s^p \to \frac{D_* \mathcal{T}^2}{L^2 \rho_*} \cdot D_s^p, \\[2mm]
\mathbf{v} \to \frac{L}{\mathcal{T}} \mathbf{v}, \quad p \to \frac{L^2 \rho_*}{\mathcal{T}^2} \cdot p, \quad \eta \to \frac{L^2 \rho_*}{\mathcal{T}} \cdot \mathbf{v}, \quad \gamma_s \to \frac{D_* e}{k_B T_*} \cdot \gamma_s, \\[2mm]
\mathbf{i}_{\xi_k} \to \frac{K_* D_* e\mathcal{E}}{K_B T_*} \cdot \mathbf{i}_{\xi_k}, \quad \varepsilon \to \varepsilon_0 \cdot \varepsilon, \quad \mathbf{g} \to \frac{L}{\mathcal{T}^2} \cdot \mathbf{g}, \quad \mathbf{j} \to \frac{K_* D_* e^2 \mathcal{E} N_A}{k_B T_*} \cdot \mathbf{j}, \\[2mm]
\chi \to \frac{L^2 \rho_* c_p}{\mathcal{T}} \cdot \chi, \quad r \to r \cdot \frac{e^2 N_A K_* \mathcal{E}^2 D_*}{\rho_* k_B T_*}, \quad h_s \to \frac{K_*}{c_p T_*} \times \\[2mm]
\times (h_* m_s N_A) \cdot h_s.
\end{array}\right\} \tag{3.1}$$

Here L, $\mathcal{T}$, $\mathcal{E}$, T_*, ρ_*, D_*, K_*, $(h_* m_s N_A)$ are, respectively, the characteristic length, time, electric field intensity, temperature, density, diffusion coefficient, molar concentration, and molar heats of the chemical reactions. For simplicity, we assume that the specific heat of the mixture c_p and the dielectric constant of water ε_0 are constant.

The characteristic time is conveniently chosen as

$$\mathscr{T} = \frac{k_B T_* L}{D_* e \mathscr{E}}.\tag{3.2}$$

Introduce the dimensionless parameters

$$\mu = \frac{k_B T_*}{e L \mathscr{E}}, \quad \theta = \frac{\varepsilon_* \varepsilon_0 \mathscr{E}}{K_* L e N_A}, \quad \delta = \frac{e N_A K_* \mathscr{E} L}{T_* c_p \rho_*}, \quad \delta_0 = \frac{k_B^2 T_*^2 N_A K_* L}{D_*^2 e \mathscr{E} \rho_*}.\tag{3.3}$$

In dimensionless variables the system of equations describing the behavior of the mixture in local chemical equilibrium has the form

$$\frac{d\rho}{dt} + \rho \operatorname{div} \mathbf{v} = 0,\tag{3.4}$$

$$\rho \frac{d\mathbf{v}}{dt} = -\nabla p + \operatorname{div}\left\{ v\left(\frac{\partial v^\alpha}{\partial x^\beta} + \frac{\partial v^\beta}{\partial x^\alpha}\right) - \frac{2}{3} v \delta_{\alpha\beta} \operatorname{div} \mathbf{v}\right\} + \rho \mathbf{g}$$
$$- \frac{1}{2} \delta_0 \theta \, (\nabla \varepsilon) \, \mathbf{E}^2 + \delta_0 \rho_{\varepsilon_0}^\varepsilon \mathbf{E},\tag{3.5}$$

$$\rho \frac{d\xi_k}{dt} + \operatorname{div} \mathbf{i}_{\xi_k} = 0, \quad k = 0, \ldots, N - R - 1,\tag{3.6}$$

$$\mathbf{i}_{\xi_k} = \rho \left\{ - \mu \nabla \left(D_{\xi_k} \xi_k\right) + \gamma_{\xi_k} \xi_k \mathbf{E} - \mu D_{\xi_k}^T \xi_k \nabla T - \mu D_{\xi_k}^p \xi_k \nabla p, \right.$$
$$k = 1, \ldots, N - R - 1,\tag{3.7}$$

$$\mathbf{j} = \mathbf{i}_{\xi_0} = \rho \sum_{k=0}^{N-R-1} \{ -\mu \nabla (\sigma_{\xi_k}^D \xi_k) + \sigma_{\xi_k} \xi_k \mathbf{E} - \mu \sigma_{\xi_k}^p \xi_k \nabla p - \mu \sigma_{\xi_k}^T \xi_k \nabla T\},\tag{3.8}$$

$$\theta \operatorname{div} (\varepsilon \mathbf{E}) = \rho \varepsilon_0, \quad \operatorname{rot} \mathbf{E} = 0,\tag{3.9}$$

$$\rho \frac{dT}{dt} + \operatorname{div} \mathbf{q} = \delta \left(\rho r + \mathbf{i}_{\xi_0} \cdot \mathbf{E}\right) + \cdots,\tag{3.10}$$

$$\mathbf{q} = -\chi \nabla T.\tag{3.11}$$

Here

$$D_{\xi_k} = \sum_{s \in I_k} D_s \beta_k^s, \quad \gamma_{\xi_k} = \sum_{s \in I_k} \gamma_s \beta_k^s\tag{3.12}$$

$$D_{\xi_k}^T = \sum_{s \in I_k} D_s^T \beta_k^s, \quad D_{\xi_k}^p = \sum_{s \in I_k} D_s^p \beta_k^s, \quad k = 1, \ldots, N-R-1, \quad (3.13)$$

$$\sigma_{\xi_k}^D = \sum_{s \in I_{0k}} \left(D_s \frac{\beta_k^s}{\alpha_s^k} z_s \right), \quad \sigma_{\xi_k} = \sum_{s \in I_{0k}} \left(\gamma_s \frac{\beta_k^s}{\alpha_s^k} z_s \right),$$

$$\sigma_{\xi_k}^p = \sum_{s \in I_{0k}} \left(D_s^p \frac{\beta_k^s}{\alpha_s^k} z_s \right), \quad \sigma_{\xi_k}^T = \sum_{s \in I_{0k}} \left(D_s^T \frac{\beta_k^s}{\alpha_s^k} z_s \right) \qquad (3.14)$$

$$k = 0, \ldots, N-R-1,$$

$$\left.\begin{aligned} &\varepsilon = 1 + \sum_{s=1}^{N} \frac{\varepsilon_s - \varepsilon_0}{\varepsilon_0} \frac{m_s N_A K_*}{\rho_*} \varphi_s(\xi_0, \ldots, \xi_{N-R-1}), \\ &a_s = \varphi_s(\xi_0, \ldots, \xi_{N-R-1}), \quad s = 1, \ldots, N, \end{aligned}\right\} \qquad (3.15)$$

$$\beta_k^s = \frac{\alpha_s^k \varphi_s(\xi_0, \ldots, \xi_{N-R-1})}{\xi_k}, \quad k = 0, \ldots, N-R-1, \quad s \in I_k. \quad (3.16)$$

In this case, the functions φ_s are determined from the system of equations (1.22)-(1.24). The dependences of $\rho, \chi, \nu, r, \varepsilon_s$ ($s = 1, \ldots, N$) on $T, \xi_0, \ldots, \xi_{N-R-1}$ are assumed to be given. In Eq. (3.10) we omit the terms describing heat evolution as a result of the chemical reactions.

For dilute solutions, the Einstein relations furnish an accurate relation between the diffusion and mobility coefficients for the charged components. In dimensionless variables these are

$$\gamma_s = z_s D_s \quad s \in I_0. \qquad (3.17)$$

Here we may write the expressions for the mobility and the molar conductivity of the chemical subsystems in the form

$$\gamma_{\xi_k} = \sum_{s \in I_k} z_s D_s \beta_k^s, \quad k = 1, \ldots, N-R-1, \qquad (3.18)$$

$$\sigma_{\xi_k} = \sum_{s \in I_{0k}} (z_s)^2 D_s \frac{\beta_k^s}{\alpha_s^k} \geq 0, \quad k = 0, \ldots, N-R-1. \qquad (3.19)$$

4. Acidity of the Solution

In this section, we present (3.4)-(3.16) in another form. Recall that one important characteristic of an aqueous solution is its acidity, the number of hydronium H_3^+O ions (or protons H^+, depending on the method used for writing the chemical reaction) in solution. The pH, specified by the relationship

$$pH = -\lg a_1, \tag{4.1}$$

(remember $[H_3^+O] = a_1$ or $[H^+] = a_1$), is a quantitative characteristic of the acidity; the molar concentration a_1 is expressed in moles/liter.

For our purposes, it is more convenient to introduce another characteristic of the acidity connected with the pH of the relationship

$$\psi = -\ln 10 \cdot (pH + \lg K_w), \tag{4.2}$$

where K_w is the dissociation constant of water expressed in moles/liter and equal to 10^{-7} mole/liter under normal conditions [12, 23].

Using (4.1) and (4.2), we obtain the molar concentration as

$$a_1 = K_w \cdot e^\psi, \quad \dim a_1 = moles/liter. \tag{4.3}$$

In dimensionless variables, this relationship has the form

$$a_1 = \frac{K_w}{K_*} e^\psi. \tag{4.4}$$

We also note that under normal conditions ($K_w = 10^{-7}$ mole/liter), $\psi > 0$ corresponds to an acid solution while $\psi < 0$ corresponds to a basic solution. For pure water $\psi = 0$.

Suppose that the original multicomponent mixture is divided into chemical subsystems, none of which contains hydronium or hydroxyl ions ($[H_3^+O] = a_1$, $[OH^-] = a_2$), i.e.,

$$1 \,\bar{\in}\, I_k, \quad 2 \,\bar{\in}\, I_k, \quad k = 1, \ldots, N - R - 1. \tag{4.5}$$

From (1.30), it follows that

$$1 \in I_{00}, \quad 2 \in I_{00}. \tag{4.6}$$

For simplicity, let us consider the case where

$$I_{00} \equiv \{1,\, 2\}. \tag{4.7}$$

Now choose the characteristic values for the molar concentration and diffusion in the form

$$K_* = 2K_w, \quad D_* = \sqrt{D_1 D_2}, \tag{4.8}$$

and introduce the dimensionless molar charges of the chemical subsystems as

$$e_{\xi_k} \equiv \sum_{s \in I_{0k}} z_s \beta_k^s, \quad k = 1,\, \ldots,\, N - R - 1. \tag{4.9}$$

Then the molar charge of the solution, (1.31), in dimensionless variables is written using (1.16), (1.21), (1.30), (3.1), and (4.7)-(4.9) as

$$\xi_0 = \operatorname{sh} \psi + \sum_{k=1}^{N-R-1} e_{\xi_k} \xi_k. \tag{4.10}$$

In an analogous fashion we obtain an expression for the electric current density

$$\mathbf{j} = (-\mu \nabla \psi + \mathbf{E}) \operatorname{ch}(\psi - \psi_*) + \sum_{k=1}^{N-R-1} \{-\mu \nabla (\sigma^D_{\xi_k} \xi_k)$$
$$+ \sigma_{\xi_k} \xi_k \mathbf{E} - \mu \sigma^T_{\xi_k} \xi_k \nabla T - \mu \sigma^p_{\xi_k} \xi_k \nabla p\}, \tag{4.11}$$

where

$$\psi_* = \frac{1}{2} \ln \frac{D_2}{D_1}. \tag{4.12}$$

The closed system of equations for describing multicomponent mixtures obtained for this case [compare with (3.4)-(3.16)] is

$$\frac{d\rho}{dt} + \rho \operatorname{div} \mathbf{v} = 0, \tag{4.13}$$

$$\rho \frac{d\mathbf{v}}{dt} = -\nabla p + \operatorname{div} \left\{ \nu \left(\frac{\partial v^\alpha}{\partial x_\beta} + \frac{\partial v^\beta}{\partial x_\alpha} \right) - \frac{2}{3} \nu \delta_{\alpha\beta} \operatorname{div} \mathbf{v} \right\} + \rho \mathbf{g}$$
$$- \frac{1}{2} \delta_0 \theta (\nabla \varepsilon) \mathbf{E}^2 + \delta_0 \rho \xi_0 \mathbf{E}, \tag{4.14}$$

$$\rho \frac{d\xi_k}{dt} + \operatorname{div} \mathbf{i}_{\xi_k} = 0. \tag{4.15}$$

$$\mathbf{i}_{\xi_k} = \rho \{-\mu \nabla (D_{\xi_k} \xi_k) + \gamma_{\xi_k} \xi_k \mathbf{E} - \mu D^T_{\xi_k} \xi_k \nabla T - \mu D^p_{\xi_k} \xi_k \nabla p\},$$
$$k = 1,\, \ldots,\, N - R - 1, \tag{4.16}$$

$$\theta \operatorname{div}(\varepsilon E) = \rho \left\{ \operatorname{sh} \psi + \sum_{k=1}^{N-R-1} e_{\xi_k} \xi_k \right\}, \qquad (4.17)$$

$$\theta \frac{\partial (\varepsilon E)}{\partial t} + \rho \left\{ \operatorname{sh} \psi + \sum_{k=1}^{N-R-1} e_{\xi_k} \xi_k \right\} v + j = J, \qquad (4.18)$$

$$j = (-\mu \nabla \psi + E) \operatorname{ch}(\psi - \psi_*) + \sum_{k=1}^{N-R-1} \{-\mu \nabla (\sigma_{\xi_k}^D \xi_k) + \sigma_{\xi_k} \xi_k E$$
$$- \mu \sigma_{\xi_k}^T \xi_k \nabla T - \mu \sigma_{\xi_k}^P \xi_k \nabla p \}, \qquad (4.19)$$

$$\rho \frac{dT}{dt} + \operatorname{div} q = \delta (\rho r + j \cdot E) + \cdots, \qquad (4.20)$$

$$q = -\chi \nabla T. \qquad (4.21)$$

The quantities $D_{\xi_k}, \gamma_{\xi_k}, D_{\xi_k}^T, D_{\xi_k}^P, \sigma_{\xi_k}^D, \sigma_{\xi_k}, \sigma_{\xi_k}^T, \sigma_{\xi_k}^P, e_{\xi_k}, \varepsilon_{\xi_k}$ are determined by the expressions (3.12)-(3.14), (3.18), and (3.19). Then

$$\beta_k^s = \frac{\alpha_s^k \varphi_s (\xi_1, \ldots, \xi_{N-R-1}, \psi)}{\xi_k}, \quad k = 1, \ldots, N-R-1, \quad s \in I_k, \quad (4.22)$$

$$a_s = \varphi_s (\xi_1, \ldots, \xi_{N-R-1}, \psi), \quad s = 1, 2, \ldots, N-R-1. \quad (4.23)$$

In this case,

$$\varphi_1 = \frac{1}{2} e^\psi, \quad \varphi_2 = \frac{1}{2} e^{-\psi}, \qquad (4.24)$$

and the functions $\varphi_3, \varphi_4, \ldots, \varphi_{N-R-1}$ are determined from the system of equations analogous to (1.22)-(1.24)

$$\left. \begin{array}{l}
\Phi_k (a_3, \ldots, a_N; \xi_1, \ldots, \xi_{N-R-1}, \psi) \\
\quad \equiv \xi_{k-1} - \sum_{s \in I_{k-1}} \alpha_s^{k-1} a_s = 0, \\
\Phi_k (a_3, \ldots, a_N; \xi_1, \ldots, \xi_{N-R-1}, \psi) \\
\quad \equiv \sigma^{(k+R-N)} (\psi, a_3, \ldots, a_N) = 0, \\
J \equiv \dfrac{D (\Phi_2, \Phi_3, \ldots, \Phi_{N-1})}{D (a_3, a_4, \ldots, a_N)} \neq 0.
\end{array} \right\} \qquad (4.25)$$

We note that system (4.13)-(4.21), in contrast to (3.4)-(3.11), is valid only when we assume local chemical equilibrium, since in its derivation we made use of relationship (1.21) for the expression of a_2 in terms of

ψ. Furthermore, in Eq. (4.18), obtained directly from (I.14.8), we completely neglected the electric current density $\mathrm{rot} \sum_{k=0}^{N} \left\{ c_k \varkappa_k \varepsilon_* \mathbf{E} \wedge (\mathbf{v} + \mathbf{w}_k) \right\}$.

In conclusion, we define closed chemical subsystems. We will say that the mixture consists of closed chemical subsystems if the following conditions are satisfied:

$$I_k \cap I_l = \varnothing, \quad k, \; l = 1, \; 2, \; \ldots, \; N - R - 1. \tag{4.26}$$

In this case, the quantities D_{ξ_k}, γ_{ξ_k}, e_{ξ_k}, and others depend only on the quantity ψ. Thus, the closed chemical subsystems do not directly interact with one another. Interaction is taken into account using the function of the solution acidity and the electric field intensity determined from Eqs. (4.17) and (4.19).

5. Applicability of the Approximation of Local Chemical Equilibrium

In this section we refine the idea of local chemical equilibrium. Let us consider whether or not the sources of the chemical reactions can be equal to zero. In dimensionless variables

$$\left. \begin{array}{l} k_r^+ \to k_r^+ \, (K_*)^{1 - \mu_l^r - \mu_p^r} \dfrac{1}{\mathscr{T}}, \quad k_r^- \to k_r^- \, (K_*)^{1 - \nu_m^r - \nu_n^r} \dfrac{1}{\mathscr{T}}, \\[2mm] \tilde{K}_r \to \tilde{K}_r \, (K_*)^{\nu_n^r + \nu_m^r - \mu_l^r - \mu_p^r}. \end{array} \right\} \tag{5.1}$$

Taking into account (1.2), (1.5), (1.6), and (3.1), we write (I.14.3) in dimensionless variables as

$$\rho \frac{da_s}{dt} + \mathrm{div} \left\{ - \rho D_s \nabla a_s + \rho \gamma_s a_s \mathbf{E} + \cdots \right\}$$
$$= \sum_{r=1}^{R} (\nu_s^r - \mu_s^r)(- k_r^+ a_l^{\mu_l^r} a_p^{\mu_p^r} + k_r^- a_m^{\nu_m^r} a_n^{\nu_n^r}). \tag{5.2}$$

We will say that the equilibrium of the rth reaction of type (1.3) is strongly shifted to the left (to the right) if $k_r^+ \gg k_r^-$ ($k_r^+ \ll k_r^-$). Introduce the symbols

$$\zeta_r = \max(k_r^+, \ k_r^-),$$

$$\tilde{\sigma}_r = \begin{cases} \left(-a_l^{\mu_l'} a_p^{\mu_p'} + \dfrac{1}{\tilde{K}_r} a_m^{\nu_m'} a_n^{\nu_n'} \right), & k_r^+ \gg k_r^-, \\[2ex] \left(-\tilde{K}_r a_l^{\mu_l'} a_p^{\mu_p'} + a_m^{\nu_m'} a_n^{\nu_n'} \right), & k_r^+ \ll k_r^-. \end{cases} \tag{5.3}$$

Then, if in *all* the reactions the equilibrium is shifted to the left or to the right, we write expression (5.2) in the form

$$\rho \frac{da_s}{dt} + \operatorname{div}\{-\rho D_s \nabla a_s + \rho \gamma_s a_s \mathbf{E} + \cdots\}$$

$$= \sum_{r=1}^{R} (\nu_s' - \mu_s') \zeta_r \tilde{\sigma}_r. \tag{5.4}$$

Let us set

$$\zeta = \min_{1 \leqslant r \leqslant R} (\zeta_r), \quad \left(\frac{\zeta_r}{\zeta}\right) = O(1), \quad r = 1, \ldots, R, \tag{5.5}$$

then

$$\frac{1}{\zeta} \rho \frac{da_s}{dt} + \frac{1}{\zeta} \operatorname{div}\{-\rho D_s \nabla a_s + \rho \gamma_s a_s \mathbf{E} + \cdots\}$$

$$= \sum_{r=1}^{R} \frac{\zeta_s}{\zeta} (\nu_s' - \mu_s') \tilde{\sigma}_r. \tag{5.6}$$

If $\zeta \to \infty$, then we may neglect the right-hand side of (5.6):

$$\sum_{r=1}^{R} \frac{\zeta_s}{\zeta} (\nu_s' - \mu_s') \tilde{\sigma}_r = 0. \tag{5.7}$$

If all the quantities σ_r are linearly independent, then from (5.7) it follows that

$$\tilde{\sigma}_r = 0, \quad r = 1, \ldots, R. \tag{5.8}$$

Thus, in order to apply the approximation of local chemical equilibrium, the conditions of linear independence of the quantities σ_r ($r = 1, \ldots, R$) and the condition $\zeta \to \infty$ must be satisfied.

Chapter III

ELECTROPHORESIS METHODS
AND THEIR
MATHEMATICAL MODELS

1. Electrophoresis Methods

In the literature, a large number of different electrophoresis methods are mentioned involving the use of specific supporting media (paper, gel, etc.) or a specific form of electrophoretic chamber [33, 37, 46, 48, 57]. However, the existing nomenclature reflects preferentially the viewpoint of experimenters and instrumentation manufacturers. If we consider all types of electrophoresis from the standpoint of their physicomathematical nature and adequate mathematical description, then many of the characteristic features of the way the method is performed prove to be unimportant. Thus, for example, the equations describing electrophoresis on paper, in liquids, and in gels are identical (as long as there is no convection).

Let us introduce some definitions needed to construct the mathematical models for electrophoresis which are widely used in the following development. Let us say that the kth chemical subsystem may be *separated* from the mixture of chemical subsystems if, as a result of processes occurring in the mixture in which there may be a fixed region D of the space occupied by the mixture, the concentration of the kth chemical subsystem becomes significantly greater than its concentration outside the region D. In this case, the requirement that a large part of the entire mass of the dissolved chemical subsystem be concentrated in the region D would be excessive. Let us designate as "the kth sample" the kth chemical subsystem which may be separated from the mixture. Let us designate the collection of chemical subsystems establishing such properties as electrical conductivity, pH, thermal conductivity, etc., for the entire mixture as a whole as the "*buffer mixture*" or the "*buffer*."

Let us designate as *electrophoresis* the process of separating the chemical subsystems (the samples) from the mixture under the influence of an electric field. In this case, the kth chemical subsystem has *amphoteric properties* if its mobility (both magnitude and sign) depends on the pH of the buffer, and a pH value exists at which the mobility is equal to zero. From the standpoint of chemical reactions, this means that the kth chemical subsystem includes ions of opposite signs, and the ratio of the concentrations for different ions depends on pH. Mathematically, the condition that the kth chemical subsystem have amphoteric properties is written as (see Sections 3 and 4 in Chapter II)

$$\gamma_{\xi_k} \equiv \gamma_{\xi_k}(\xi_1, \xi_2, \ldots, \psi) \equiv \gamma_{\xi_k}(\psi), \tag{1.1}$$

$$\gamma_{\xi_k}(\psi_k^i) \equiv 0. \tag{1.2}$$

Let us call the quantity ψ_k^i the *isoelectric value* of the kth chemical subsystem. Furthermore, let us introduce the idea of the *isoionic value* ψ_k^e for the kth chemical subsystem, at which the molar charge of the subsystem is equal to zero:

$$e_{\xi_k}(\psi_k^e) \equiv 0, \quad e_{\xi_k} \equiv e_{\xi_k}(\psi). \tag{1.3}$$

Examples of amphoteric substances are amino acids, peptides, proteins, etc.

Among all the diverse electrophoresis types we will limit ourselves to the study of isotachophoresis, isoelectric focusing, and zone electrophoresis. Let us list the most important distinguishing features and the regions of application for these types of electrophoresis for fractionation of biopolymers.

In *isotachophoresis* all the samples are simultaneously components of the buffer, and all the solutes of the original multicomponent mixture should have a common ion. Isotachophoresis is based on the familiar moving boundary method (see [46]). A solution containing ions of the mixture to be separated is placed between two indicator electrolytes, the first of which is called the leading electrolyte and the last of which is called the terminal electrolyte (the terminator). When the transport rates (see below) for all the ions of the solution are different (faster than for the leading electrolyte and slower than for the terminator), a steady state is established upon migration of the ions: the mixture is divided into zones which move at the same velocity. Isotachophoresis is one of the most effective electromigration methods in chemistry and biology (see, for example, [46, 47, 60, 76, 95], and also papers given in the bibliographic index of the LKB company, Acta Isotachophoretica – 456 citations for the period from 1967 to 1980). Among the most important applications of isotachophoresis we note the separation of inorganic ions (in particular isotopes with close mobilities), strong and weak acids and their salts, and nucleotides. Isotachophoresis is especially widely used for preparative (in gram amounts) purification of proteins; in this case, it is possible to achieve very high (hundred-fold) concentration of the sample of interest to us in the "pure" zone, which is impossible in other types of electrophoresis. Capillary isotachophoresis is most frequently used, in gels or on cellulose acetate plates.

In *isoelectric focusing* a pH gradient is created using a buffer, the samples have amphoteric properties, and their concentrations as a rule are significantly less than the concentrations of the buffer components. The history of the method and the means of creating pH gradients are described in the introduction to Chapters VI and VII (also see, for example, [51, 74, 77, 91, 92]). In recent years isoelectric focusing (IEF) has undergone very active development and has found wide application [62]: in clinical chemistry, especially for analysis of blood proteins [49, 50]; in immunology, for example, in estimating the synthesis rates for different immunoglobulins; in oncology, for early diagnostics, for example, according to the level of specific enzymes; in membranology for obtaining and analyzing the spectrum of membrane proteins; in chemotaxonomy, especially when morphological methods are difficult or impossible; in agriculture for the identification of types of agricultural plants; in microbiology, forensic medicine, the food industry, etc. In such applications an extremely high resolution is achieved – up to 0.005 pH unit – which under certain conditions may correspond to replacement of only one amino acid in the protein.

Here we should note that the increasing resolution of all the electromigration methods for fractionation has complicated the investigation of macromolecules in biology: in light of these methods, almost each biologically active molecule represents a family of macromolecules consisting of "charge isomers," "size isomers," or products of reaction with small or macromolecular reagents. Several representatives of these families of macromolecules may have biological significance; the rest may represent an artifact of the separation or chemical conversion products. This makes identification of the molecules and their comparative analysis difficult; more accurate ways of characterizing the results of electrophoresis are required than estimating them in terms of "two zones," "three zones," "fast," "slow," etc.

In *zone electrophoresis* a constant pH value is maintained in the mixture using a buffer, at which pH the mobilities of the different samples are different from one another, and accordingly their separation occurs during the electrophoresis process. Usually the concentrations of all the samples are less than the concentrations of the buffer components. One of the most important applications of zone electrophoresis is fractionation of nucleic acids and their fragments, which in the overwhelming majority of cases is carried out in gels – polyacrylamide and agarose (see, for example,

[37]). Nucleic acids always have a negative overall electric charge, the magnitude of which depends little on the pH of the medium. Since the charge-to-mass ratio is practically the same for all nucleic acids, as a rule their separation in electrophoresis is due to the difference in sizes and shape of the molecules and not to difference in charges.

The reaction between nucleic acids and gels has been insufficiently studied up to the present time: the functional relationships between mobility and relative molecular mass used are empirical in character [37, 61]. Apparently we need a more detailed investigation of the mechanism of charge formation for a nucleic acid in aqueous solution [96]. We should note that electrophoresis has proven to be an extremely effective means for determining fine conformational differences in nucleic acids [72, 73, 79, 80]. For example, using electrophoresis, the number of superhelices has been determined in circular superhelical DNA [70, 81]; i.e., the method makes it possible to specifically detect differences in conformation, since the molecular mass does not change upon spiralization or despiralization of DNA.

Electrophoresis of nucleic acids has occupied a key position in the technology of genetic engineering, where with the use of electrophoresis restriction fragments of DNA have been identified and their sizes have been determined; the combination of treatment with enzymes and labeled preparations together with electrophoresis in polyacrylamide gel and autoradiography has become the most effective method for determining the primary structure of nucleic acids [37], which is especially important in connection with advances in biological and chemical gene synthesis.

The possibilities for zone electrophoresis of proteins are diverse; this is also, as a rule, carried out in gels [33, 37, 48, 74]. Using this method, not only have the mobilities and molecular masses of proteins been determined [83, 94], but also, for example, association constants [86]. This method occupies an important position in population genetics as a means for studying genetic variability [100].

We should emphasize one more time that the theory developed in this book refers to free liquid electrophoresis, and may be extended to electrophoresis in gels only if chemical reactions do not occur between the gels and the biopolymers.

The differential equations describing zone electrophoresis and isoelectric focusing are the same. There is a finer difference between these two methods. From the mathematical standpoint, zone electrophoresis is described by solutions for "short" times: in every case, the times for carrying out the process should not be greater than the time it takes for the fastest zone to pass through the entire column, and it should also be shorter than the time in which diffusion blurs the already separated zones.

As far as isoelectric focusing is concerned, here the required state is established for "long" times. The end result of this process is a steady state in which each sample has collected in the neighborhood of its isoelectric point; the width of the neighborhood is determined by the diffusion rate. In this state, those samples are separated in which the indicated neighborhoods do not intersect. Thus, in contrast to zone electrophoresis, isoelectric focusing can counteract diffusion.

2. Additional Simplifications

The reductions of the general equations for the dynamics of multi-component mixtures, carried out in Chapter II by means of separation of the basic processes into slow and fast processes, are still insufficient for constructing practically useful mathematical models of electrophoresis for biopolymers that allow for discernable solutions. For this purpose we need to take one more step: specific allowance for the chemistry of the mixture under consideration and also possibly allowance for the specific characteristics of the experiment under consideration, letting us simplify the model by discarding processes which weakly affect the electrophoresis. Later in this chapter we construct several models having apparently a rather limited region of application, but allowing us to arrive at a final result which allows comparison with experiment. We hope that the examples given below have a certain methodological value, illustrating the approach to constructing mathematical models for physicochemical processes in multicomponent reactive mixtures subject to the influence of electric fields, temperature fields, etc. Obviously the models presented in the best possible case reflect only the most important aspects of the phenomena under consideration and ultimately require further improvement. In this direction, we attach primary importance to accounting for the temperature dependences of the material constants (the mobility, diffusion coefficients, etc.) and also to dropping the condition that the solution be weak by going from the concentrations of the substances to their activities. In Chapter V (Section 3), we consider an example in which we show that taking into account the temperature dependence of the mobility allows us to explain the curvature of the zones in zone electrophoresis.

At the basis of the construction of mathematical models for electrophoresis is the system of equations (II.3.4)-(II.3.11) or its modification (II.4.13)-(II.4.21). These systems of equations are not closed as long as the coefficients D_{ξ_k}, γ_{ξ_k}, $D_{\xi_k}^T$ are not determined. Our goal is to consider some examples of chemically reactive mixtures found in local chemical equilibrium that would most completely correspond to the requirements forming the basis of the definition of the different types of electrophoresis. Knowing the specific chemical reactions allows us (using the algorithm described in Chapter II) to determine the form of the coefficients D_{ξ_k}, γ_{ξ_k}, and thus to close the system of equations.

In order to simplify the presentation, let us restrict ourselves to calculation of the degrees of dissociation β_k^s for specific mixtures, and let us write only the transport equations for the chemical subsystems and the equations for the electric field (II.3.6)-(II.3.9), omitting the rest of the equations of the complete system (consequently, of the modified system of equations). Furthermore, let us give the specific expressions only for the coefficients D_{ξ_k}, γ_{ξ_k}, $\alpha_{\xi_k}^D$, α_{ξ_k}; and in the expressions for the fluxes (II.3.7)-(II.3.8), we will omit terms containing the coefficients $D_{\xi_k}^T$, $D_{\xi_k}^P$, $\alpha_{\xi_k}^D$, $\alpha_{\xi_k}^T$. In this case, the complete system of equations, taking into account all the terms, may be easily restored since the quantities β_k^s will be given.

The systems of equations written below will be complete if we make the following assumptions. The mixture as a whole is found in mechanical equilibrium; there are no barodiffusion or thermal diffusion effects; the dielectric constant of the mixture is equal to the dielectric constant of the solvent; the diffusion coefficients and mobilities for the individual components of the original mixture do not depend on temperature (the latter means that the thermal conductivity equation is separated from the system and may be integrated independently). The following relationships correspond to these conditions:

$$\mathbf{v} = 0, \quad \rho = 1, \quad \varepsilon = 1, \quad D_s^T = D_s^p = 0, \quad s = 1, \ldots, N,$$

$$(2.1)$$

$$\frac{\partial}{\partial T} D_s \equiv \frac{\partial}{\partial T} \gamma_s \equiv 0, \quad s = 1, \ldots, N,$$

$$\frac{d}{dt} = \frac{\partial}{\partial t}. \tag{2.2}$$

The condition $\rho = 1$ corresponds to the fact that the density of the solution is chosen as the quantity ρ_* in (II.3.1) and (II.3.3). Of course, going from $\partial/\partial t$ to d/dt when $\rho \neq$ const is not difficult. Furthermore, we will assume that the Einstein relations are satisfied (II.3.17):

$$\gamma_s = z_s D_s, \quad s \in l_0. \tag{2.3}$$

3. Simplest Model for Isoelectric Focusing and Zone Electrophoresis: One-Component Buffer, One Sample

As has been noted, the equations describing isoelectric focusing and zone electrophoresis are the same. Let us illustrate this in the following example. Let us consider a multicomponent solution in which dissociation of a monobasic acid HA occurs (for example, H_3BO_3, $A^- \equiv H_2BO_3^-$) and some amino acid $NH_3^+RCOO^-$, where R is the amino acid residue (for example, $R \equiv CH_2$ for glycine). In aqueous solution, the dissociation reactions occur according to the scheme (see, for example, [23, 27])

$$\left. \begin{aligned} & NH_3^+RCOOH \rightleftarrows NH_3^+RCOO^- + H^+, \\ & NH_3^+RCOO^- \rightleftarrows NH_2RCOO^- + H^+, \\ & HA \rightleftarrows H^+ + A^-. \end{aligned} \right\} \tag{3.1}$$

Let us introduce the following symbols ([ ] means molar concentration):

$$a_1 = [H^+], \quad a_2 = [HA], \quad a_3 = [A^-], \quad a_4 = [NH_3^+RCOO^-], \tag{3.2}$$
$$a_5 = [NH_2RCOO^-], \quad a_6 = [NH_3^+RCOOH].$$

Obviously, the charge of the components will be

$$z_1 = z_6 = 1, \quad z_3 = z_5 = -1, \quad z_2 = z_4 = 0. \tag{3.3}$$

In the new symbols, scheme (3.1) has the form [compare with (II.1.3)]

$$\left. \begin{aligned} & r = 1) \; a_6 \rightleftarrows a_4 + a_1, \quad r = 2) \; a_4 \rightleftarrows a_5 + a_1, \\ & r = 3) \; a_2 \rightleftarrows a_1 + a_3, \quad R = 3, \quad N = 6. \end{aligned} \right\} \tag{3.4}$$

All the following is valid for *any* seven-component mixture whose component charges have the values (3.3) and whose dissociation reactions proceed according to scheme (3.4).

For the rth reaction in (3.4), let us introduce sources of the form (II.1.5)

$$\left. \begin{aligned} & \sigma^{(1)} = -k_1^+ a_6 + k_1^- a_1 a_4, \quad \sigma^{(2)} = -k_2^+ a_4 + k_2^- a_1 a_5, \\ & \sigma^{(3)} = -k_3^+ a_2 + k_3^- a_1 a_3. \end{aligned} \right\} \tag{3.5}$$

Then the chemical kinetics equations for molar concentrations a_s ($s = 1, ..., 6$) have the form [see (II.1.1), (II.1.2), (II.1.6)]

$$\left. \begin{aligned} & \frac{\partial a_1}{\partial t} = \sigma^{(1)} + \sigma^{(2)} + \sigma^{(3)}, \quad \frac{\partial a_2}{\partial t} = -\sigma^{(3)}, \quad \frac{\partial a_3}{\partial t} = \sigma^{(3)}, \\ & \frac{\partial a_4}{\partial t} = \sigma^{(1)} - \sigma^{(2)}, \quad \frac{\partial a_5}{\partial t} = \sigma^{(2)}, \quad \frac{\partial a_6}{\partial t} = -\sigma^{(1)}. \end{aligned} \right\} \tag{3.6}$$

After establishment of chemical equilibria, the original seven-component mixture is converted to a *two-component* mixture of closed chemical subsystems with concentrations [see (II.1.9)-(II.1.11)]

$$\xi_2 = a_3 + a_2, \quad I_2 = \{2, 3\},$$

$$\xi_1 = a_4 + a_5 + a_6, \quad I_1 = \{4, 5, 6\}, \quad (I_1 \cap I_2 = \varnothing). \tag{3.7}$$

The molar charge of the mixture of chemical subsystems is

$$\xi_0 = a_1 + a_6 - a_3 - a_5, \quad I_0 = \{1, 3, 5, 6\}. \tag{3.8}$$

We note that for a solution in which reactions (3.1) proceed, ξ_1 is the analytic concentration of the acid HA and ξ_2 is the analytic concentration of the amino acid $NH_3^+RCOO^-$. In accordance with (II.1.30), let us introduce the set of indices

$$I_{01} = I_0 \cap I_1 = \{5, 6\}, \quad I_{02} = (I_0 \setminus I_{01}) \cap I_2 = \{3\}, \left. \right\}$$
$$I_{00} = ((I_0 \setminus I_{01}) \setminus I_{02}) \cap I_0 = \{1\}. \left. \right\} \tag{3.9}$$

Let us determine the degree of dissociation for the components of the chemical subsystems

$$a_2 = \beta_2^2 \xi_2, \quad a_3 = \beta_2^3 \xi_2, \quad \beta_2^2 + \beta_2^3 = 1,$$

$$a_4 = \beta_1^4 \xi_1, \quad a_5 = \beta_1^5 \xi_1, \quad a_6 = \beta_1^6 \xi_1, \quad \beta_1^4 + \beta_1^5 + \beta_1^6 = 1. \tag{3.10}$$

Using (3.5), we write the conditions for local chemical equilibrium $\sigma^{(r)} \equiv 0$, $r = 1, 2, 3$ [see (II.1.18), (II.1.20)],

$$\frac{a_1 a_4}{a_6} = K_1, \quad \frac{a_1 a_5}{a_4} = K_2, \quad \frac{a_1 a_3}{a_2} = K_3, \tag{3.11}$$

where K_r are the reduced dissociation constants. For reactions (3.1): K_1 is the dissociation constant for the carboxyl group; K_2 is the dissociation constant for the amino group; K_3 is the dissociation constant for the acid HA (in the case of a polybasic acid, the first-stage dissociation constant).

Substituting (3.10) into (3.11), we obtain the equations for determining the degrees of dissociation β_k^s:

$$\frac{\beta_2^3}{1 - \beta_2^3} = \frac{K_3}{a_1}, \tag{3.12}$$

$$\frac{1 - \beta_1^5 - \beta_1^6}{\beta_1^6} = \frac{K_1}{a_1}, \quad \frac{\beta_1^5}{1 - \beta_1^5 - \beta_1^6} = \frac{K_2}{a_1}. \tag{3.13}$$

Let us introduce the symbols

$$a_1 = K_* e^\psi, \quad \alpha = \frac{K_3}{K_*}, \quad \psi_1^e = \frac{1}{2} \ln \frac{K_1 K_2}{K_*^2}, \quad m_1 = \frac{1}{2} \sqrt{\frac{K_1}{K_2}},$$

$$K_* = K_w, \quad \dim(K_i, K_*, K_w) = \frac{\text{moles}}{\text{m}^3}, \quad D = \sqrt{D_5 D_6}, \quad (3.14)$$

$$\psi_1^i = \psi_1^e + \delta_1, \quad \delta_1 = \frac{1}{2} \ln \frac{D_5}{D_6}, \quad D_* = D_{1,\,\dim}.$$

Solving (3.12) and (3.13), we obtain

$$\beta_2^3 = \frac{\alpha}{\alpha + e^\psi}, \quad \beta_2^2 = \frac{e^\psi}{\alpha + e^\psi}, \quad \beta_1^5 = \frac{\frac{1}{2} e^{-(\psi - \varphi_1^e)}}{\operatorname{ch}(\psi - \psi_1^e) + m_1},$$

$$\beta_1^6 = \frac{\frac{1}{2} e^{(\psi - \psi_1^e)}}{\operatorname{ch}(\psi - \psi_1^e) + m_1}, \quad \beta_1^4 = \frac{m_1}{\operatorname{ch}(\psi - \psi_1^e) + m_1}. \quad (3.15)$$

Substituting (3.15) into (II.3.12), (II.3.18), (II.3.19), and (II.4.9), and taking into account (3.3), (3.7)-(3.10), and (3.14), we obtain expressions for the diffusion coefficients, mobilities, diffusion conductivities, conductivities, and molar charges of the components for the mixture of chemical subsystems:

$$D_{\xi_2} = D_2 \frac{e^\psi}{\alpha + e^\psi} + D_3 \frac{\alpha}{\alpha + e^\psi} > 0,$$

$$\gamma_{\xi_2} = -D_3 \frac{\alpha}{\alpha + e^\psi}, \quad e_{\xi_2} = -\frac{\alpha}{\alpha + e^\psi}, \quad (3.16)$$

$$\sigma_{\xi_2}^D = -D_3 \frac{\alpha}{\alpha + e^\psi}, \quad \sigma_{\xi_2} = D_3 \frac{\alpha}{\alpha + e^\psi} > 0,$$

$$D_{\xi_1} = \frac{m_1 D_4 + D \operatorname{ch}(\psi - \psi_1^i)}{\operatorname{ch}(\psi - \psi_1^e) + m_1} > 0, \quad \gamma_{\xi_1} = D \frac{\operatorname{sh}(\psi - \psi_1^i)}{\operatorname{ch}(\psi - \psi_1^e) + m_i},$$

$$e_{\xi_1} = \frac{\operatorname{sh}(\psi - \psi_1^e)}{\operatorname{ch}(\psi - \psi_1^e) + m_1}, \quad \sigma_{\xi_1}^D = \gamma_{\xi_1}, \quad \sigma_{\xi_1} = D \frac{\operatorname{ch}(\psi - \psi_1^e)}{\operatorname{ch}(\psi - \psi_1^e) + m_1} > 0. \quad (3.17)$$

By definition [see (1.1), (1.2)] the chemical subsystem with concentration ξ_1 has amphoteric properties: there exists the isoelectric value $\psi_1{}^i$ such that

$$\gamma_{\xi_1}(\psi_1^i) \equiv 0. \quad (3.18)$$

The quantity $\psi_1{}^e$ is the isoionic value.

Substituting (3.16) and (3.17) into (II.3.6)-(II.3.9), and taking into account (2.1) and (3.14), we obtain equations describing the behavior of a two-component mixture of chemical subsystems:

$$\frac{\partial \xi_2}{\partial t} + \text{div}\left\{-\mu\nabla\left(\frac{\alpha D_3 + D_2 e^\psi}{\alpha + e^\psi}\xi_2\right) - D_3\frac{\alpha}{\alpha + e^\psi}\xi_2 E\right\} = 0, \qquad (3.19)$$

$$\frac{\partial \xi_1}{\partial t} + \text{div}\left\{-\mu\nabla\left[\frac{m_1 D_4 + D\,\text{ch}\,(\psi - \psi_1^f)}{\text{ch}\,(\psi - \psi_1^f) + m_1}\right] + D\frac{\text{sh}\,(\psi - \psi_1^f)}{\text{ch}\,(\psi - \psi_1^f) + m_1}\xi_1 E\right\} = 0,$$
$$(3.20)$$

$$\frac{\partial \xi_0}{\partial t} + \text{div}\,\mathbf{j} = 0, \qquad (3.21)$$

$$\xi_0 = e^\psi - \frac{\alpha}{\alpha + e^\psi}\xi_2 + \frac{\text{sh}\,(\psi - \psi_1^e)}{\text{ch}\,(\psi - \psi_1^e) + m_2}\cdot\xi_1, \qquad (3.22)$$

$$\mathbf{j} = -\mu e^\psi\nabla\psi + \mu\nabla\left(D_3\frac{\alpha}{\alpha + e^\psi}\xi_2\right) - \mu\nabla\left[\frac{D\,\text{sh}\,(\psi - \psi_1^f)}{\text{ch}\,(\psi - \psi_1^f) + m_1}\xi_1\right]$$

$$+ \mathbf{E}\left\{e^\psi + D_3\frac{\alpha}{\alpha + e^\psi}\xi_2 + D\frac{\text{ch}\,(\psi - \psi_1^f)}{\text{ch}\,(\psi - \psi_1^e) + m_1}\xi_1\right\}, \qquad (3.23)$$

$$\theta\,\text{div}\,\mathbf{E} = \xi_0, \quad \text{rot}\,\mathbf{E} = 0. \qquad (3.24)$$

The system (3.19)-(3.24) is closed. Of course, we must add initial and boundary conditions to this system. This system is suitable both for describing zone electrophoresis and for describing isoelectric focusing. We should consider the chemical subsystem with the concentration ξ_2 as the buffer, and the chemical subsystem with the concentration ξ_1 as the sample. Specifying special initial distributions of the concentration ξ_2 in the mixture, we may specify the pH gradient, or, conversely, we may make the pH value constant. It should not confuse the reader that, in this model, the pH value may change in the course of electrophoresis. The condition for determining the relative ratio of concentrations of buffer and sample is not satisfied for both types of electrophoresis; furthermore, there is one sample (i.e., there is nothing to fractionate). As will be shown later on, these contradictions are easily eliminated.

4. Model with One-Component Buffer and Several Samples

If the mixture contains more than one sample, the model given in the preceding section is not made significantly more complicated. Let us consider a solution in which dissociation of the monobasic acid HA and the set of amino acids $NH_3^+R_kCOO^-$ ($k = 1, ..., M$) occurs, where R_k are the amino acid residues, for example: $R_1 \equiv CH_2$ (glycine), $R_2 \equiv CH_3CH$ (α-alanine), etc. For simplicity, let us assume that the amino acids are capable of dissociation only relative to one carboxyl and one amino group. The case where the amino acids each contain several amino and carboxyl groups is considered in Section 6. Furthermore, often we cannot take into account the dissociation of several amino and carboxyl groups for a sufficiently large interval of variation in pH. Let us consider the following dissociation scheme in solution:

$$\left.\begin{aligned} &NH_3^+ R_k COOH \rightleftarrows NH_3^+ R_k COO^- + H^+, \\ &NH_3^+ R_k COO^- \rightleftarrows NH_2 R_k COO^- + H^+, \quad k = 1, \ldots, M, \\ &HA \rightleftarrows H^+ + A^- \end{aligned}\right\} \quad (4.1)$$

Let us introduce the symbols [compare with (3.2) and (3.3)]

$$\left.\begin{aligned} &a_1 = [H^+], \quad a_2 = [HA], \quad a_3 = [A^-], \\ &a_{3k+1} = [NH_3^+ R_k COO^-], \quad a_{3k+2} = [NH_2 R_k COO^-], \\ &a_{3k+3} = [NH_3^+ R_k COOH], \quad k = 1, \ldots, M \end{aligned}\right\} \quad (4.2)$$

$$\left.\begin{aligned} &z_1 = 1, \quad z_2 = 0, \quad z_3 = -1, \\ &z_{3k+1} = 0, \quad z_{3k+2} = -1, \quad z_{3k+3} = 1, \quad k = 1, \ldots, M. \end{aligned}\right\} \quad (4.3)$$

Then scheme (4.1) takes on the form [compare with (3.4)]

$$\left.\begin{aligned} &r = 2k - 1) \; a_{3k+3} \rightleftarrows a_{3k+1} + a_1, \\ &r = 2k) \; a_{3k+1} \rightleftarrows a_{3k+2} + a_1, \quad k = 1, \ldots, M, \\ &r = 2M + 1) \; a_2 \rightleftarrows a_1 + a_3, \quad N = 3M + 3, \quad R = 2M + 1. \end{aligned}\right\} \quad (4.4)$$

Let us write the chemical kinetics equations as [compare with (3.6)]

$$\left.\begin{aligned} &\frac{\partial a_1}{\partial t} = \sum_{r=1}^{2M+1} \sigma^{(r)}, \quad \frac{\partial a_2}{\partial t} = -\sigma^{(2M+1)}, \quad \frac{\partial a_3}{\partial t} = \sigma^{(2M+1)}, \\ &\frac{\partial a_{3k+1}}{\partial z} = \sigma^{(2k-1)} - \sigma^{(2k)}, \quad \frac{\partial a_{3k+2}}{\partial t} = \sigma^{(2k)}, \\ &\frac{\partial a_{3k+3}}{\partial t} = -\sigma^{(2k-1)}, \quad k = 1, \ldots M. \end{aligned}\right\} \quad (4.5)$$

The original $(3M + 4)$-component mixture, after establishment of local chemical equilibrium, is converted to a $(M + 1)$-component mixture of closed chemical subsystems with concentrations

$$\begin{aligned} \xi_k &= a_{3k+1} + a_{3k+2} + a_{3k+3}, \quad k = 1, \ldots M, \\ \xi_{M+1} &= a_2 + a_3 \end{aligned} \quad (4.6)$$

and molar charge for the mixture

$$\xi_0 = a_1 - a_3 + \sum_{k=1}^{M} (a_{3k+3} - a_{3k+2}). \quad (4.7)$$

Let us write the relationships for the degrees of dissociation for the components of the chemical subsystems:

$$
\left.\begin{array}{l}
a_2 = \beta_{M+1}^2 \xi_{M+1}, \quad a_3 = \beta_{M+1}^3 \xi_{M+1}, \quad \beta_{M+1}^2 + \beta_{M+1}^3 = 1, \\[2mm]
a_{3k+1} = \beta_k^{3k+1} \xi_k, \quad a_{3k+2} = \beta_k^{3k+2} \xi_k, \quad a_{3k+3} = \beta_k^{3k+3} \xi_k, \\[2mm]
\sum_{i=1}^{3} \beta_k^{3k+i} = 1.
\end{array}\right\} \tag{4.8}
$$

Let us introduce the symbols

$$
\left.\begin{array}{l}
a_1 = K_* e^\psi, \quad \alpha = \dfrac{K_{2M+1}}{K_*}, \quad \psi_p^e = \dfrac{1}{2}\ln\dfrac{K_{2p-1}K_{2p}}{K_*^2}, \\[3mm]
\overline{D}_p = \sqrt{D_{3p+2}D_{3p+3}}, \quad \psi_p^i = \psi_p^e + \delta_p, \quad \delta_p = \dfrac{1}{2}\ln\dfrac{D_{3p+2}}{D_{3p+3}}, \\[3mm]
p = 1, \ldots, M; \\[2mm]
D_* = D_{1,\,\mathrm{dim}}, \quad K_* = K_w, \quad \dim\{K_p,\, K_*,\, K_w\} = \dfrac{\mathrm{moles}}{\mathrm{m}^3}.
\end{array}\right\} \tag{4.9}
$$

The systems of equations for determining β_ks are obtained from the conditions for local chemical equilibrium:

$$
\left.\begin{array}{l}
\dfrac{\beta_{M+1}^3}{1 - \beta_{M+1}^3} = \dfrac{K_{2M+1}}{a_1}, \quad \dfrac{1 - \beta_k^{3k+2} - \beta_k^{3k+3}}{\beta_k^{3k+3}} = \dfrac{K_{2k-1}}{a_1}, \\[4mm]
\dfrac{\beta_k^{3k+2}}{1 - \beta_k^{3k+2} - \beta_k^{3k+3}} = K_{2k}, \quad k = 1, \ldots, M.
\end{array}\right\} \tag{4.10}
$$

Comparing (4.10) with (3.12) and (3.13), in analogy with Section 3, we obtain expressions for β_ks, and then also for $D_{\xi_k}, \gamma_{\xi_k}, \ldots,$:

$$
\left.\begin{array}{l}
D_{\xi_{M+1}} = D_2 \dfrac{e^\psi}{\alpha + e^\psi} + D_3 \dfrac{\alpha}{\alpha + e^\psi} > 0, \\[3mm]
\gamma_{\xi_{M+1}} = -D_3 \dfrac{\alpha}{\alpha + e^\psi}, \quad e_{\xi_{M+1}} = -\dfrac{\alpha}{\alpha + e^\psi}, \\[3mm]
\sigma_{\xi_{M+1}}^D = \gamma_{\xi_{M+1}}, \quad \sigma_{\xi_{M+1}} = -\gamma_{\xi_{M+1}},
\end{array}\right\} \tag{4.11}
$$

$$
\left.\begin{array}{l}
D_{\xi_k} = \dfrac{m_k D_{3k+1} + \overline{D}_k \operatorname{ch}(\psi - \psi_k^i)}{\operatorname{ch}(\psi - \psi_k^e) + m_k} > 0, \\[3mm]
\sigma_{\xi_k}^D \equiv \gamma_{\xi_k} = \overline{D}\dfrac{\operatorname{sh}(\psi - \psi_k^i)}{\operatorname{ch}(\psi - \psi_k^e) + m_k}, \quad e_{\xi_k} = \dfrac{\operatorname{sh}(\psi - \psi_k^e)}{\operatorname{ch}(\psi - \psi_k^e) + m_k}, \\[3mm]
\sigma_{\xi_k} = \overline{D}\dfrac{\operatorname{ch}(\psi - \psi_k^i)}{\operatorname{ch}(\psi - \psi_k^e) + m_k} > 0, \quad k = 1, \ldots, M.
\end{array}\right\} \tag{4.12}
$$

Obviously, the following conditions are satisfied:

$$\gamma_{\xi_k}(\psi_k^i) \equiv 0, \quad e_{\xi_k}(\psi_k^e) \equiv 0, \quad k = 1, \ldots, M. \qquad (4.13)$$

Thus, ψ_k^i, ψ_k^e are the isoelectric and isoionic values for the kth amphoteric chemical subsystem (see Section 1 of this chapter). Finally, we write the equations describing the processes of zone electrophoresis and isoelectric focusing in a $(M + 1)$-component mixture of chemical subsystems [compare with (3.19)-(3.24):

$$\frac{\partial \xi_k}{\partial t} + \operatorname{div} \left\{ \mu \nabla \frac{(m_k D_{3k+1} + \overline{D}_k) \operatorname{ch}(\psi - \psi_k^i)}{\operatorname{ch}(\psi - \psi_k^e) + m_k} \xi_k \right.$$

$$\left. + \overline{D}_k \frac{\operatorname{sh}(\psi - \psi_k^i)}{\operatorname{ch}(\psi - \psi_k^e) + m_k} \xi_k \mathbf{E} \right\} = 0, \qquad (4.14)$$

$$\frac{\partial \xi_{M+1}}{\partial t} + \operatorname{div} \left\{ -\mu \nabla \left(\frac{\alpha D_3 + D_2 e^{\psi}}{\alpha + e^{\psi}} \xi_{M+1} \right) - D_3 \frac{\alpha}{\alpha + e^{\psi}} \xi_{M+1} \mathbf{E} \right\} = 0, \quad (4.15)$$

$$\frac{\partial \xi_0}{\partial t} + \operatorname{div} \mathbf{j} = 0, \qquad (4.16)$$

$$\xi_0 = e^{\psi} - \frac{\alpha}{\alpha + e^{\psi}} \xi_{M+1} + \sum_{k=1}^{M} \frac{\operatorname{sh}(\psi - \psi_k^e)}{\operatorname{ch}(\psi - \psi_k^e) + m_k} \xi_k, \qquad (4.17)$$

$$\mathbf{j} = -\mu e^{\psi} \nabla \psi + \mu \nabla \left(D_3 \frac{\alpha}{\alpha + e^{\psi}} \xi_{M+1} \right) - \mu \sum_{k=1}^{M} \nabla \left[\frac{\overline{D}_k \operatorname{sh}(\psi - \psi_k^i)}{\operatorname{ch}(\psi - \psi_k^e) + m_k} \xi_k \right]$$

$$+ \mathbf{E} \left\{ e^{\psi} + D_3 \frac{\alpha}{\alpha + e^{\psi}} \xi_{M+1} + \sum_{k=1}^{M} \overline{D}_k \frac{\operatorname{ch}(\psi - \psi_k^i)}{\operatorname{ch}(\psi - \psi_k^e) + m_k} \xi_k \right\}, \qquad (4.18)$$

$$\theta \operatorname{div} \mathbf{E} = \xi_0, \quad \operatorname{rot} \mathbf{E} = 0. \qquad (4.19)$$

The system obtained should be supplemented by initial and boundary conditions.

Let us note an important characteristic feature of the electrophoresis models obtained: allowance for reactions between components of the mixture is made only through the acidity ψ and the electric field intensity $\mathbf{E}$, characterizing the properties of the mixture as a whole.

5. Simplified Models for the Case of Weak Electrolytes

The mathematical models constructed in the preceding sections may be even more simplified by assuming that the buffer concentration is significantly greater than the sample concentration, and thus the mathematical model is brought closer to the situation which is most frequently realized in electrophoresis practice. For the sake of concreteness, we will concern ourselves with the model constructed in Section 3.

Let us consider the case when the buffer is acid, i.e., $K_w \gg K_3$; $\alpha = K_3/K_w \ll 1$. In this case, in the system of equations (3.19)-(3.24), the natural small parameter α appears. (If the buffer is basic, i.e., $K_w \ll K_3$, of course we choose as the parameter $\alpha = K_w/K_3 \ll 1$.) We make the following assumptions:

$$\xi_1 = O(1), \quad \xi_2 = O\left(\frac{1}{\alpha}\right), \quad \alpha \to 0. \tag{5.1}$$

$$m_1 = \frac{m}{\alpha}, \quad m = O(1), \quad \alpha \to 0 \left(\frac{K_1}{K_w} = O\left(\frac{1}{\alpha}\right), \frac{K_2}{K_w} = O(\alpha), \alpha \to 0\right), \tag{5.2}$$

$$D_s = O(1), \quad s = 1, \ldots, 6; \tag{5.3}$$

$$\psi_1^s - \psi_1^e = O(\alpha), \quad \alpha \to 0 \ (D_5 - D_6 = O(\alpha), D - D_4 = O(\alpha), \alpha \to 0), \tag{5.4}$$

$$E = O\left(\frac{1}{\alpha}\right), \quad j = O\left(\frac{1}{\alpha}\right), \quad \alpha \to 0, \tag{5.5}$$

$$\theta = \theta_0 \alpha^2, \quad \theta_0 = O(1), \quad \alpha \to 0. \tag{5.6}$$

Condition (5.1) means that the buffer concentration is much greater than the sample concentration. Assumption (5.2) corresponds to the fact that the degrees of dissociation for the charged components are small:

$$\beta_2^3 = O(\alpha), \quad \beta_2^2 = 1 + O(\alpha), \quad \beta_1^5 = O(\alpha),$$
$$\beta_1^6 = O(\alpha), \quad \beta_1^4 = 1 + O(\alpha), \quad \alpha \to 0. \tag{5.7}$$

Thus, the solution under consideration is a weak electrolyte. In this case, the concentration of the neutral components in the original solution differs little from the concentrations of the components in the mixture of chemical subsystems:

$$a_2 = \xi_2 + O(\alpha), \quad a_4 = \xi_1 + O(\alpha), \quad \alpha \to 0. \tag{5.8}$$

Condition (5.3) means that, in the solution under consideration, the diffusion coefficients for the components are comparable to one another. Condition (5.4) requires that there be little difference between the isoelectric value and the isoionic value. It is automatically satisfied if the mobilities of the differently charged components of the chemical subsystem differ little in magnitude, which is a rather reasonable assumption when the mass of the amino acid residue R_k is much higher than the mass of the H^+ ions separated upon dissociation. [Let us recall that if the Einstein relations (2.3) are satisfied for singly charged ions, in dimensionless variables the magnitudes of the mobilities coincide with the magnitudes of the diffusion coefficients.] The requirement $D - D_4 = O(\alpha)$ is not very important and is assumed for convenience, although by virtue of what has been noted above, it is a rather reasonable assumption. Condition (5.5) in essence allows us to neglect the contribution of diffusion to the electric current in the mixture. In this case, however, the diffusion contributions to the transport processes of the components are retained. Finally, condition (5.6) corresponds to a solution which is close to electrically neutral. Making the replacements

$$\xi_2 \to \frac{1}{\alpha}\,\xi_2, \quad E \to \frac{1}{\alpha}\,E, \quad j \to \frac{1}{\alpha}\,j, \quad \psi_i^t \equiv \psi^t, \tag{5.9}$$

we will look for a solution to system (3.19)-(3.24) in the form of series in powers of α:

$$\xi_i = \xi_i^{(0)} + \alpha\xi_i^{(1)} + \cdots, \quad E = E^{(0)} + E^{(1)}\alpha + \cdots,$$

$$\psi = \psi^{(0)} + \psi^{(1)}\alpha + \cdots. \tag{5.10}$$

Substituting (5.10) into (3.19)-(3.24) and restricting ourselves to terms of order $O(1)$, we obtain (for convenience we omit the superscripts)

$$\frac{\partial \xi_1}{\partial t} + \mathrm{div}\left\{-\mu D_4 \nabla \xi_1 + \frac{D}{m}\,\mathrm{sh}\,(\psi - \psi^t)\,\xi_1 E\right\} = 0, \tag{5.11}$$

$$\frac{\partial \xi_2}{\partial t} + \mathrm{div}\left\{-\mu D_2 \nabla \xi_2 - D_3\,\frac{\xi_2}{e^\psi}\,E\right\} = 0, \tag{5.12}$$

$$\mathrm{div}\,j = 0; \quad \mathrm{rot}\,E = 0, \tag{5.13}$$

$$e^\psi - \frac{\xi_2}{e^\psi} = 0, \tag{5.14}$$

$$j = \left(e^\psi + D_3\,\frac{\xi_2}{e^\psi}\right) E. \tag{5.15}$$

Expressing e^ψ from (5.14) and substituting it into (5.12) and (5.15), finally we have, taking into account (5.13),

$$\frac{\partial \xi_1}{\partial t} + \mathrm{div}\,\{-\mu D_4 \nabla \xi_1 + \beta e^{-\psi}\,\mathrm{sh}\,(\psi - \psi^t)\,\xi_1\} = 0, \tag{5.16}$$

TABLE 1. Molar Charge of Glycine Calculated
from Formulas (4.12) and (5.21)

pH	ψ	e_{ξ_1}	$e_{\xi_1}^{weak}$
8.737	−4.0	−8.308·10^{-2}	−9.061·10^{-2}
8.303	−3.0	−3.225·10^{-2}	−3.333·10^{-2}
7.000	0	−1.634·10^{-3}	−1.637·10^{-3}
6.783	0.5	−9.686·10^{-4}	−9.697·10^{-4}
6.566	1.0	−5.493·10^{-4}	−5.497·10^{-4}
6.348	1.5	−2.698·10^{-4}	−2.700·10^{-4}
6.131	2.0	−5.916·10^{-5}	−5.918·10^{-5}
6.065	2.153	0	0
5.914	2.5	1.364·10^{-4}	1.365·10^{-4}
5.697	3.0	3.668·10^{-4}	3.670·10^{-4}
3.526	8.0	6.256·10^{-2}	6.673·10^{-2}
·2.657	10.0	3.303·10^{-1}	4.931·10^{-1}

$$\frac{\partial \xi_2}{\partial t} + \text{div} \left\{ -\mu D_2 \nabla \xi_2 \right\} = 0, \tag{5.17}$$

$$e^{\psi} = \sqrt{\xi_2}, \quad \mathbf{E} = \frac{e^{-\psi}}{1+D_3} \mathbf{j} \equiv \frac{1}{\sqrt{\xi_2}\,(1+D_3)} \mathbf{j}, \tag{5.18}$$

$$\text{div}\, \mathbf{j} = 0, \quad \text{rot}\, \mathbf{j} = \nabla \psi \wedge \mathbf{j}, \tag{5.19}$$

where

$$\beta = \frac{D}{m} \frac{\mathbf{j}}{1+D_3}. \tag{5.20}$$

System (5.16)-(5.20) completely satisfies the requirements formulated in Section 1 for isoelectric focusing. The gradient in the quantity ψ (the pH gradient) may be specified using the initial concentration distribution of the buffer ξ_2. In this case, the change in the pH gradient is determined only by the diffusion process of the buffer. The sample with concentration ξ_1 has amphoteric properties, and its concentration is significantly less than the concentration of the buffer. Obviously, this system is also appropriate for describing zone electrophoresis. Specifying the buffer concentration $\xi_2 = $ const, we thus maintain a constant value $\psi \neq \psi^i$; in this case, the mobility is $\gamma_{\xi_1} = $ const.

The model constructed in this section may be called a diffusion-free model; there is no inverse effect of the sample on the buffer and the field. The system of equations (5.16)-(5.20) is split: first we determine the solution to system (5.17)-(5.19), i.e., we determine the characteristics of the mixture as a whole specified by the buffer; then we integrate (5.16), determining the motion of the sample, which does not exert any effect on the buffer. An analogous construction may be transferred without difficulty to the model given in Section 4.

In conclusion, let us introduce the results of calculations of the molar charge e_{ξ_1} according to formula (4.12) and the quantity $e_{\xi_1}^{\text{weak}}$, obtained in the weak electrolyte approximation:

$$e_{\xi_1}^{\text{weak}} = \frac{\alpha \, \text{sh} \, (\psi - \psi_1^e)}{m}.$$ (5.21)

Let us choose the following values for the parameters:

$$\psi_1^e = 2.152916, \quad K_3/K_w = 5.7544 \cdot 10^{-3}, \quad K_w = 10^{-7} \text{ mole/liter},$$
$$K_1 = 10^{-2,35} \text{ mole/liter}, \quad K_2 = 10^{-9.78} \text{ mole/liter}, \quad K_3 = 10^{-9.24} \text{ mole/liter},$$
$$m = 2593.999, \quad pI = 6.065.$$

The constants K_1 and K_2 correspond to glycine; K_3 corresponds to boric acid. The results of the calculations are given in Table 1.

Comparison of the values of e_{ξ_1} and $e_{\xi_1}^{\text{weak}}$ show that in the interval pH $\in$ [4, 8] for glycine we may, to a sufficient degree of accuracy, use the weak electrolyte approximation. Of course, it is easy to verify that the quantities γ_{ξ_1}, $\gamma_{\xi_1}^{\text{weak}}$ and D_{ξ_1}, $D_{\xi_1}^{\text{weak}}$, etc., in the indicated pH interval also differ little from one another. In presenting the table, it should furthermore be recalled that the weak electrolyte model has a limited region of application. Obviously, in the example under consideration the weak electrolyte model is inapplicable when the pH is outside the limits of the range [4, 8].

6. Mobility and Molar Charge of an Amino Acid with Several Carboxyl and Amino Groups

Let us consider dissociation in aqueous solution of amino acids which have several amino or carboxyl groups capable of ionization. As is shown in Sections 3 and 4, the components of the amino acid formed upon dissociation make up a closed chemical subsystem. In this case, the degrees of dissociation for the components of the subsystem depend only on the concentration of the ions a_1 (H$^+$). An analogous situation occurs in the case of amino acids with several amino or carboxyl groups. Let us illustrate this by an example, considering the dissociation of an amino acid having two amino groups (NH$_2$)$_2$$R$COOH, where R is the amino acid residue. For example, $R \equiv CH(CH_2)_4$ for lysine, which is a "basic" amino acid. The dissociation of such an amino acid in aqueous solution proceeds according to the scheme

$$\left. \begin{aligned} &NH_3^+NH_3^+RCOOH \rightleftarrows NH_3^+NH_3^+RCOOH^- + H^+, \\ &NH_3^+NH_3^+RCOO^- \rightleftarrows NH_3^+NH_2RCOO^- + H^+, \\ &NH_3^+NH_2RCOO^- \rightleftarrows NH_2NH_2RCOO^- + H^+. \end{aligned} \right\} \qquad (6.1)$$

Introducing the symbols

$$a_1 = [H^+], \quad a_2 = [NH_3^+NH_2RCOO^-], \quad a_3 = [NH_3^+NH_3^+RCOO^-],$$
$$a_4 = [NH_3^+NH_3^+RCOOH], \quad a_5 = [NH_2NH_2RCOO^-], \qquad (6.2)$$

$$z_1 = 1, \quad z_2 = 0, \quad z_3 = 1, \quad z_4 = 2, \quad z_5 = -1, \qquad (6.3)$$

we write

$$\left. \begin{aligned} &\cdots\cdots\cdots\cdots\cdots \\ r &= 1) \; a_4 \rightleftarrows a_3 + a_1, \\ r &= 2) \; a_3 \rightleftarrows a_2 + a_1, \\ r &= 3) \; a_2 \rightleftarrows a_5 + a_1. \\ &\cdots\cdots\cdots\cdots\cdots \end{aligned} \right\} \qquad (6.4)$$

Then the chemical kinetics equations have the form

$$\left. \begin{aligned} &\cdots\cdots\cdots\cdots \\ &\frac{\partial a_2}{\partial t} = \sigma^{(2)} - \sigma^{(3)} \\ &\frac{\partial a_3}{\partial t} = \sigma^{(1)} - \sigma^{(2)} \\ &\frac{\partial a_4}{\partial t} = -\sigma^{(1)} \\ &\frac{\partial a_5}{\partial t} = \sigma^{(3)} \\ &\cdots\cdots\cdots\cdots \end{aligned} \right\} \qquad (6.5)$$

where $\sigma^{(r)} \equiv 0$, $r = 1, 2, 3$, are the sources of the corresponding reactions.

Obviously, the components a_2, a_3, a_4, and a_5 form a chemical subsystem:

$$\xi_1 = a_2 + a_3 + a_4 + a_5. \qquad (6.6)$$

TABLE 2. Molar Charge of Lysine e_{ξ_1} for Different pH Values, Calculated Using Formulas (6.11) and (6.14)

pH	Formula (6.11)	Formula (6.14)
0	1.993	1.000
1	1.938	1.000
2	1.602	1.000
3	1.131	1.000
4	1.015	1.000
5	1.001	$9.999 \cdot 10^{-1}$
6	$9.990 \cdot 10^{-1}$	$9.989 \cdot 10^{-1}$
7	$9.889 \cdot 10^{-1}$	$9.889 \cdot 10^{-1}$
8	$8.985 \cdot 10^{-1}$	$8.985 \cdot 10^{-1}$
9	$4.486 \cdot 10^{-1}$	$8.617 \cdot 10^{-1}$
9.74	$1.922 \cdot 10^{-7}$	0
10	$-1.488 \cdot 10^{-1}$	$-1.488 \cdot 10^{-1}$
11	$-7.430 \cdot 10^{-1}$	$-7.430 \cdot 10^{-1}$
12	$-9.672 \cdot 10^{-1}$	$-9.672 \cdot 10^{-1}$
13	$-9.966 \cdot 10^{-1}$	$-9.966 \cdot 10^{-1}$
14	$-9.996 \cdot 10^{-1}$	$-9.996 \cdot 10^{-1}$

Introducing the degrees of dissociation

$$a_2 = \beta_1^2 \xi_1 \equiv (1 - \beta_1^3 - \beta_1^4 - \beta_1^5) \xi_1, \quad a_3 = \beta_1^3 \xi_1,$$

$$a_4 = \beta_1^4 \xi_1, \quad a_5 = \beta_1^5 \xi_1, \tag{6.7}$$

from the conditions of local chemical equilibrium ($\sigma^{(r)} \equiv 0$, $r = 1, 2, 3$), we obtain the system of equations

$$\frac{\beta_1^3}{\beta_1^4} = \frac{K_1}{a_1}, \quad \frac{1 - \beta_1^3 - \beta_1^4 - \beta_1^5}{\beta_1^3} = \frac{K_2}{a_1}, \quad \frac{\beta_1^5}{1 - \beta_1^3 - \beta_1^4 - \beta_1^5} = \frac{K_3}{a_1}, \tag{6.8}$$

where K_1 is the reduced dissociation constant for the carboxyl group; K_2, K_3 are the reduced dissociation constants for the first and second amino groups. Solving system (6.8), we obtain

$$\beta_1^3 = \frac{\frac{K_1}{a_1}}{\Delta (a_1)}, \quad \beta_1^4 = \frac{1}{\Delta (a_1)}, \quad \beta_1^5 = \frac{\frac{K_1 K_2 K_3}{a_1^3}}{\Delta (a_1)},$$

$$\Delta (a_1) \equiv 1 + \frac{K_1}{a_1} + \frac{K_1 K_2}{a_1^2} + \frac{K_1 K_2 K_3}{a_1^3}. \tag{6.9}$$

For the mobility and the molar charge of the chemical subsystem, using (II.3.18), (II.4.9), and (6.3) we derive

$$\gamma_{\xi_1} = \frac{D_3 \frac{K_1}{a_1} + 2D_4 - D_5 \frac{K_1 K_2 K_3}{a_1^3}}{\Delta (a_1)}, \tag{6.10}$$

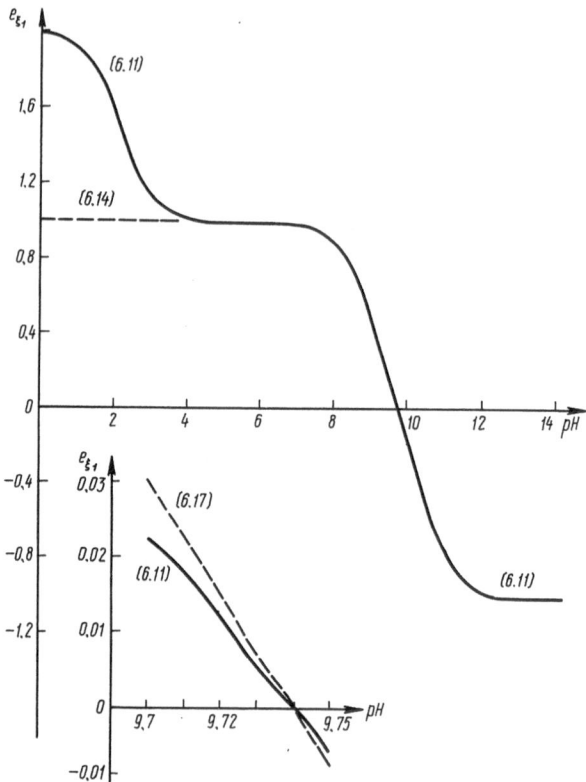

Fig. 4. Theoretical titration curve for lysine. In parentheses we indicate the numbers of the formulas given in the text corresponding to the curve.

$$e_{\xi_1} = \frac{\dfrac{K_1}{a_1} + 2 - \dfrac{K_1 K_2 K_3}{a_1^3}}{\Delta(a_1)}. \tag{6.11}$$

Let us consider the case where the following relationships are satisfied:

$$\frac{K_1}{K_w} = O\left(\frac{1}{\alpha}\right), \quad \frac{K_2}{K_w} = O(1),$$

$$\frac{K_3}{K_w} = O(1), \tag{6.12}$$

$$a_1 = e^{\psi} \cdot K_w, \quad e^{\psi} = O(1), \quad \alpha \to 0,$$

TABLE 3. Molar Charge of Lysine e_{ξ_1}, Calculated Using Formulas (6.11), (6.14), and (6.17)

pH	Formula (6.11)	Formula (6.14)	Formula (6.17)
9.50	$1{,}370839 \cdot 10^{-1}$	$1{,}370838 \cdot 10^{-1}$	$1.885127 \cdot 10^{-1}$
9.60	$7{,}930922 \cdot 10^{-2}$	$7{,}930913 \cdot 10^{-2}$	$1{,}063822 \cdot 10^{-1}$
9.70	$2{,}256645 \cdot 10^{-2}$	$2{,}256631 \cdot 10^{-2}$	$2{,}991703 \cdot 10^{-2}$
9.72	$1{,}127994 \cdot 10^{-2}$	$1{,}127995 \cdot 10^{-2}$	$1.494262 \cdot 10^{-2}$
9.73	$5{,}639621 \cdot 10^{-3}$	$5{,}639564 \cdot 10^{-3}$	$7.469308 \cdot 10^{-3}$
9.74	$1{,}922512 \cdot 10^{-7}$	0	0
9.75	$-5{,}639483 \cdot 10^{-3}$	$-5{,}639637 \cdot 10^{-3}$	$-7.469406 \cdot 10^{-3}$

where $\alpha \to 0$ is a small parameter introduced in analogy to Sections 3 and 4.

Using the symbols

$$\psi^e = \frac{1}{2} \ln \frac{K_2 K_3}{K_w^2}, \quad \psi^i = \psi^e + \delta, \quad \delta = \frac{1}{2} \ln \frac{D_5}{D_3},$$

$$D = \sqrt{D_3 D_5}, \quad m = \frac{1}{2} \sqrt{\frac{K_2}{K_3}} = O(1) \tag{6.13}$$

and rejecting terms of higher order in α, we obtain from (6.10) and (6.11)

$$\gamma_{\xi_1} = \frac{\text{sh}(\psi - \psi^i)}{\text{ch}(\psi - \psi^e) + m}, \quad e_{\xi_1} = \frac{\text{sh}(\psi - \psi^e)}{\text{ch}(\psi - \psi^e) + m}. \tag{6.14}$$

[Compare the expressions obtained with (3.17) and (4.12).]

Thus, under certain conditions the mathematical models of Sections 3 and 4 are not substantially changed, even if amino acids capable of ionization to several amino or carboxyl groups exist in the mixture. Of course, in this case we must recall that the condition $e^\psi \equiv O(1)$ is of decisive importance, since it is obvious that

$$\gamma_{\xi_1} \to 2D_4, \quad e_{\xi_1} \to 2, \quad \text{when } e^\psi \to \infty \quad (\psi \gg 0), \tag{6.15}$$

$$\gamma_{\xi_1} \to -D_5, \quad e_{\xi_1} \to -1, \quad e^\psi \to 0 \quad (\psi \ll 0) \tag{6.16}$$

independently of the order of magnitude of K_r/K_w ($r = 1, 2, 3$). Cases are possible in which the quantities γ_{ξ_1}, e_{ξ_1} may be replaced as follows:

$$\gamma_{\xi_1} = \frac{D\alpha}{m_*}\,\mathrm{sh}\,(\psi - \psi^i), \quad e_{\xi_1} = \frac{\alpha}{m_*}\,\mathrm{sh}\,(\psi - \psi^e), \quad \frac{m_*}{\alpha} = m. \quad (6.17)$$

For this, it is sufficient to require that the following relationships be satisfied:

$$\frac{K_1}{K_w} = O\left(\frac{1}{\alpha}\right), \quad \frac{K_2}{K_w} = O\left(\frac{1}{\alpha}\right), \quad \frac{K_3}{K_w} = O\,(\alpha), \quad \alpha \to 0;$$

$$a_1 = K_w e^{\psi}, \quad e^{\psi} = O\,(1), \quad m_* = O\,(1), \quad \alpha \to 0. \tag{6.18}$$

Retaining the symbols in (6.13) and rejecting terms of higher order in α, from (6.10) and (6.11) we derive (6.17) (compare with Section 5). Let us illustrate this by the following example. Choosing lysine as the dissociating acid [35] with

$$K_1 = 6.606934 \cdot 10^{-3}\,\text{moles/liter}, K_2 = 1.2202 \cdot 10^{-9}\ \text{moles/liter},$$

$$K_3 = 2.951211 \cdot 10^{-11}\,\text{moles/liter}, \, pI = 9.74,$$

$$\psi_1^e = -6.309083, \quad m = 3.082976,$$

let us give the results of calculation for the quantity e_{ξ_1} from formulas (6.11), (6.14), and (6.17) (Fig. 4; Tables 2 and 3). It is quite evident that in the interval pH $\in$ [4, 14], formulas (6.11) and (6.14) practically coincide. The situation is significantly worse with formulas (6.11) and (6.17). This means that, in this case, formulas (6.17) are weakly applicable; they may be used only in the neighborhood of the isoelectric point. Of course, this is explained by the fact that conditions (6.18) are not satisfied. Let us note that the curve on Fig. 4 is the titration curve for lysine, which may also be obtained experimentally [77, 93].

7. Mathematical Model of Isoelectric Focusing and Zone Electrophoresis in the Case of a Two-Component Buffer

The primary goal of the example presented in this section is to show that, from a mathematical standpoint, use of several two-component buffers does not lead to substantial complication of the electrophoresis model. Let us consider a buffer created by some acid HA and the salt of this acid MA, for example the acetate buffer

$$HA \equiv CH_3COOH, \ MA = CH_3COONa, \ M \equiv Na, \ A^- \equiv CH_3COO^-.$$

Let us introduce the following dissociation reactions:

$$
\left.
\begin{array}{l}
r = 1)\ a_6 \rightleftarrows a_4 + a_1, \\
r = 2)\ a_4 \rightleftarrows a_5 + a_1, \\
r = 3)\ a_2 \rightleftarrows a_3 + a_1, \\
r = 4)\ a_8 \rightleftarrows a_3 + a_7, \quad N = 8, \quad R = 4.
\end{array}
\right\}
\tag{7.1}
$$

In this case,

$$
z_1 = z_6 = z_7 = 1, \quad z_2 = z_8 = z_4 = 0, \quad z_3 = z_5 = -1.
\tag{7.2}
$$

A concrete example for scheme (7.1), the dissociation of the acetate buffer, has the form

$$
\left.
\begin{array}{l}
HA \rightleftarrows A^- + H^+, \\
MA \rightleftarrows A^- + M^+, \\
R^+ \rightleftarrows R + H^+, \\
R \rightleftarrows R^- + H^+,
\end{array}
\right\}
\tag{7.3}
$$

where

$$
\left.
\begin{array}{l}
a_1 = [H^+], \quad a_2 = [HA], \quad a_3 = [A^-], \quad a_4 = [R] = \\
= [NH_3^+ RCOO^-], \quad a_5 = [R^-] = [NH_2 RCOO^-], \\
a_6 = [R^+] = [NH_3^+ RCOOH], \quad a_7 = [M^+], \quad a_8 = [MA].
\end{array}
\right\}
\tag{7.4}
$$

We rewrite the chemical kinetics equations in the form

$$
\left.
\begin{array}{l}
\dfrac{\partial a_1}{\partial t} = \sigma^{(1)} + \sigma^{(3)} + \sigma^{(4)}, \quad \dfrac{\partial a_2}{\partial t} = -\sigma^{(1)}, \quad \dfrac{\partial a_3}{\partial t} = \sigma^{(1)} + \sigma^{(2)}, \\[2mm]
\dfrac{\partial a_4}{\partial t} = \sigma^{(3)} - \sigma^{(4)}, \quad \dfrac{\partial a_5}{\partial t} = -\sigma^{(3)}, \quad \dfrac{\partial a_6}{\partial t} = -\sigma^{(3)}, \\[2mm]
\dfrac{\partial a_7}{\partial t} = \sigma^{(2)}, \quad \dfrac{\partial a_8}{\partial t} = -\sigma^{(2)},
\end{array}
\right\}
\tag{7.5}
$$

$$
\sigma^{(4)} = -k_4^+ a_8 + k_4^- a_3 a_7.
\tag{7.6}
$$

The quantities $\sigma^{(1)}$, $\sigma^{(2)}$, and $\sigma^{(3)}$ are specified by relations (3.5).

The original nine-component mixture in chemical equilibrium represents a three-component mixture of chemical subsystems with concentrations

$$\left.\begin{array}{l} \xi_2 = a_2 + a_3 + a_8, \quad I_2 = \{2,\ 3,\ 8\}, \\ \xi_1 = a_4 + a_5 + a_6, \quad I_1 = \{4,\ 5,\ 6\}, \\ \xi_3 = a_7 + a_8, \quad I_3 = \{7,\ 8\}. \end{array}\right\} \tag{7.7}$$

Furthermore, in analogy to Section 3, we write

$$\xi_0 = + a_1 + a_6 + a_7 - a_3 - a_5, \quad I_0 = \{1,\ 3,\ 5,\ 6,\ 7\}, \tag{7.8}$$

$$I_{02} = \{3\}, \quad I_{01} = \{5,\ 6\}, \quad I_{03} = \{7\}, \quad I_{00} = \{1\}. \tag{7.9}$$

Let us introduce the degrees of dissociation $\beta_k{}^s$:

$$\left.\begin{array}{l} a_2 = \beta_2^2 \xi_2, \quad a_3 = \beta_2^3 \xi_2, \quad a_8 = \beta_2^8 \xi_2 \equiv (1 - \beta_2^2 - \beta_2^3)\,\xi_2, \\ a_4 = \beta_1^4 \xi_1 \equiv (1 - \beta_1^5 - \beta_1^6)\,\xi_1, \quad a_5 = \beta_1^5 \xi_1, \quad a_6 = \beta_1^6 \xi_1, \\ a_7 = \beta_3^7 \xi_3, \quad a_8 = \beta_3^8 \xi_3 \equiv (1 - \beta_3^7)\,\xi_3. \end{array}\right\} \tag{7.10}$$

The equilibrium conditions for the chemical reactions have the form [compare with (3.12) and (3.13)]

$$\frac{\beta_2^3}{1 - \beta_2^3 - \beta_2^8} = \frac{K_3}{a_1}, \quad \frac{\xi_2 \beta_2^3 \beta_3^7}{1 - \beta_3^7} = K_4, \tag{7.11}$$

$$\frac{1 - \beta_1^5 - \beta_1^6}{\beta_1^6} = \frac{K_1}{a_1}, \quad \frac{\beta_1^5}{1 - \beta_1^5 - \beta_1^6} = \frac{K_2}{a_1}. \tag{7.12}$$

Equation (7.12) for determining $\beta_1{}^4$, $\beta_1{}^5$, $\beta_1{}^6$ coincides with (3.13), and their solution coincides with (3.15).

Let us require that the solution of the salt MA be a strong electrolyte, i.e., that it completely dissociate into ions. For this, it is sufficient to set

$$K_4 = \infty \tag{7.13}$$

Then from (7.10) and the second equation of (7.11), we derive

$$\beta_3^7 = 1, \quad \beta_2^8 = 0, \quad \beta_3^8 = 0 \quad (\xi_2 \neq 0). \tag{7.14}$$

Substituting the expressions obtained into the first equation of (7.11), we obtain the relationship coinciding with (3.12),

$$\frac{\beta_2^3}{1 - \beta_2^3} = \frac{K_3}{a_1}. \tag{7.15}$$

Thus, the quantities $\beta_2{}^3$, $\beta_2{}^2$ are determined by the relationships (3.15).

Retaining the symbols in (3.14) and requiring that relationships (5.1)-(5.6) be satisfied, in analogy to Section 5, when $\alpha \to 0$ we obtain for the principal terms of order $O(1)$ [compare with (5.16)-(5.19)]

$$\frac{\partial \xi_2}{\partial t} + \operatorname{div}\left\{- \mu D_2 \nabla \xi_2 - D_3 \frac{\xi_2}{e^\psi} \mathbf{E}\right\} = 0, \tag{7.16}$$

$$\frac{\partial \xi_1}{\partial t} + \operatorname{div}\left\{- \mu D_4 \nabla \xi_1 + \frac{D}{m} \operatorname{sh}(\psi - \psi^i) \xi_1 \mathbf{E}\right\} = 0, \tag{7.17}$$

$$\operatorname{div}(\xi_3 \mathbf{E}) = 0, \tag{7.18}$$

$$\operatorname{div} \mathbf{j} = 0, \quad \operatorname{rot} \mathbf{E} = 0, \tag{7.19}$$

$$e^\psi + \xi_3 - \frac{\xi_2}{e^\psi} = 0, \tag{7.20}$$

$$\mathbf{j} = \left(e^\psi + D_7 \xi_3 + D_3 \frac{\xi_2}{e^\psi}\right) \mathbf{E}. \tag{7.21}$$

8. Mathematical Model
of Isotachophoresis

In accordance with the requirements formulated in the definition of isotachophoresis, let us consider an aqueous solution of substances which yield a common ion upon dissociation. As an example, let us assume a salt BA_k ($k = 1, \ldots, n$) of the base B. Dissociation in aqueous solution proceeds according to the scheme

$$BA_k \rightleftarrows B^+ + A_k^-. \tag{8.1}$$

Let us introduce the symbols

$$\left.\begin{array}{l} a_1 = [B^+], \quad a_{2k} = [BA_k], \quad a_{2k+1} = [A_k^-], \\ z_1 = 1, \quad z_{2k} = 0, \quad z_{2k+1} = -1, \quad k = 1, \ldots n, \\ \sigma^{(r)} = - k_r^+ a_{2r} + k_r^- a_1 a_{2r+1}, \quad r = 1, \ldots, n. \end{array}\right\} \tag{8.2}$$

The chemical kinetics equations have the form

$$\frac{\partial a_1}{\partial t} = \sum_{r=1}^{n} \sigma^{(r)}, \tag{8.3}$$

$$\frac{\partial a_{2k}}{\partial t} = - \sigma^{(k)}, \quad \frac{\partial a_{2k+1}}{\partial t} = \sigma^{(k)}, \quad k = 1, \ldots, n. \tag{8.4}$$

The original $(2n + 2)$-component mixture in local chemical equilibrium represents an n-component mixture of chemical subsystems with concentrations

$$\xi_k = a_{2k} + a_{2k+1}, \quad I_k = \{2k, 2k + 1\}, \quad k = 1, \ldots, n \qquad (8.5)$$

and molar charge

$$\xi_0 = a_1 - \sum_{k=1}^{n} a_{2k+1}, \quad I_0 = \{1, 3, 5, \ldots 2n + 1\}. \qquad (8.6)$$

For the set of indices I_{0k}, we derive

$$I_{0k} = \{2k + 1\}, \quad k = 0, \ldots, n. \qquad (8.7)$$

Let us introduce the degrees of dissociation for the components of the chemical subsystems:

$$a_{2k} = \beta_k^{2k} \xi_k \equiv (1 - \beta_k^{2k+1}) \xi_k, \quad a_{2k+1} = \beta_k^{2k+1} \xi_k. \qquad (8.8)$$

From the chemical equilibrium conditions $(\sigma^{(k)} = 0, k = 1, \ldots, n)$ and using (8.2) and (8.8), we obtain

$$\frac{\beta_m^{2m+1}}{1 - \beta_m^{2m+1}} = \frac{K_m}{a_1}, \quad m = 1, \ldots, n, \qquad (8.9)$$

from which we derive

$$\beta_m^{2m+1} = \frac{K_m}{K_m + a_1}, \quad \beta_m^{2m} = \frac{a_1}{K_m + a_1}, \quad m = 1, \ldots, n, \qquad (8.10)$$

where K_m is the reduced dissociation constant for the salt BA_m $(m = 1, \ldots, n)$.

Let us assume that the mixture of chemical subsystems is close to electrical neutrality, i.e.,

$$\theta \rightarrow 0. \qquad (8.11)$$

Then from (II.3.6)-(II.3.9), taking into account (2.1)-(2.3) and (II.3.12), we obtain the system of equations describing the isotachophoresis process:

$$\frac{\partial \xi_i}{\partial t} + \operatorname{div} \left\{ -\mu \nabla \left(\frac{D_{2i+1} K_i + D_{2i} a_1}{K_i + a_1} \xi_i \right) - \frac{K_i D_{2i+1} \xi_i}{K_i + a_1} E \right\} = 0, \quad i = 1, \ldots, n,$$

$$(8.12)$$

$$a_1 = \sum_{i=1}^{n} \frac{\dot{K}_i}{K_i + a_1} \, \xi_i, \tag{8.13}$$

$$\mathbf{j} = -\mu \nabla a_1 + \mu \nabla \left(\sum_{i=1}^{n} \frac{D_{2i+1} K_i}{K_i + a_1} \, \xi_i \right) + \left(a_1 + \sum_{i=1}^{n} \frac{D_{2i+1} K_i}{K_i + a_1} \, \xi_i \right) \mathbf{E}, \tag{8.14}$$

$$\operatorname{div} \mathbf{j} = 0, \quad \operatorname{rot} \mathbf{E} = 0. \tag{8.15}$$

Let us consider this system in the case of no diffusion, i.e.,

$$\mu = 0. \tag{8.16}$$

Let us introduce the new unknown functions v_i ($i = 1, \dots, n$) using the relations

$$\xi_i = \frac{K_i + a_1}{K_i} \, v_i. \tag{8.17}$$

Then system (8.12)-(8.14) is reduced to the form

$$\frac{\partial}{\partial t} \left\{ v_i + \frac{1}{K_i} \sum_{s=1}^{n} v_s \cdot v_i \right\} - \operatorname{div} \left\{ \frac{\mathbf{j} D_{2i+1} v_i}{\sum_{s=1}^{n} (1 + D_{2s+1}) \, v_s} \right\} = 0, \quad i = 1, \dots, n, \tag{8.18}$$

$$a_1 = \sum_{i=1}^{n} v_i, \tag{8.19}$$

$$\mathbf{E} = \frac{1}{\sum_{i=1}^{n} (1 + D_{2i+1}) \, v_i} \, \mathbf{j}, \tag{8.20}$$

$$\operatorname{div} \mathbf{j} = 0, \quad \operatorname{rot} \mathbf{E} = 0. \tag{8.21}$$

The model obtained is the diffusion-free model for isotachophoresis in arbitrary solutions. Further simplifications of this system for weak and strong electrolytes are given in Chapter IV.

9. Boundary Conditions for Models of Electrophoresis

In this section, we give several examples of boundary conditions for different electrophoresis models. Let us consider the region D occupied by the mixture with boundaries $\Gamma_0 + \Gamma$. Let us introduce the electric potential φ

$$E = - \nabla \varphi \quad (\text{rot } E \equiv 0). \tag{9.1}$$

We will call the boundary Γ_0 a solid wall if $\varphi(x, t) \equiv 0$, $x \in \Gamma_0$, and we will call the boundary Γ the electrode if $\varphi(x, t) \neq 0$, $x \in \Gamma$. In electrophoresis practice, the region D frequently represents a cylinder; and thus the boundary Γ is the lateral surface of the cylinder and the boundary Γ is the base of the cylinder. In the following, we will call such a region the electrophoretic column. Classification of the boundary of the region D according to the value of the potential φ on the boundary means, of course, that the boundary Γ_0 is assumed to be incapable of adsorbing the charged components of the mixture, which might change its potential.

For all the components of the multicomponent mixture of chemical subsystems, at the boundary Γ_0 we will specify the boundary impermeability condition: the normal component of the molar flux of the component is equal to zero:

$$\Gamma_0 : i_{\xi_k} \cdot n_0 = 0, \quad k = 1, 2, \dots, \tag{9.2}$$

$$i_{\xi_k} = - \mu \nabla (D_k \xi_k) + \gamma_k \xi_k E, \quad k = 1, 2, \dots. \tag{9.3}$$

Here, n_0 is the outward normal to the boundary Γ_0. A consequence of this condition is the relationship

$$\Gamma_0 : j \cdot n_0 \equiv j_{n_0} = 0, \tag{9.4}$$

meaning that the normal component of the electric current density j_{n_0} at the boundary Γ_0 is equal to zero; there is no electric current through Γ_0.

For the components of the buffer at the electrodes, we will specify the following boundary condition:

$$\Gamma : \xi_k |_\Gamma = \tilde{\xi}_k (t), \quad k \in I_\delta, \tag{9.5}$$

where I_δ is the set of indices for the buffer components, $\tilde{\xi}_k(t)$ is the specified concentration of the kth component of the buffer on the boundary. Condition (9.5) corresponds to the fact that we do not take into account electrode reactions at the electrodes Γ. In electrophoresis practice, such a condition is realized by separation of the electrode region, where the electrode reactions occur, from the basic electrophoretic chamber containing the multicomponent mixture (for example, using gel plugs). In this case, the buffer concentration at the boundary is chosen so that we may ignore the reaction products at the electrodes.

If, as a result of electrode reactions, a large number of buffer ions are generated which significantly increase the current density in the solution, in our opinion a reasonable boundary condition is the requirement of impermeability of the boundary Γ for the neutral components of the buffer. Using (II.1.28) and (I.14.5) and taking into account (2.1), this condition is written in the form

$$\Gamma : i_s \cdot n = 0, \qquad (9.6)$$

$$i_s = -\mu D_s \nabla \left(\frac{\beta_k^s}{\alpha_s^k} \xi_k \right), \qquad k \in I_\delta, \quad s \in I_k, \qquad (9.7)$$

where n is the outward normal to the boundary Γ, and i_s is the flux of the neutral component of the buffer. We note that, in the case of the simplified models for weak electrolytes, condition (9.6) has the form, to an accuracy up to order of magnitude $O(\alpha^2)$ [see Section 5 and (5.8)],

$$\Gamma : \nabla \xi_k \cdot n = 0, \quad k \in I_\delta. \qquad (9.8)$$

For the components of the mixture which are the samples, the boundary impermeability condition is reasonable:

$$\Gamma : i_{\xi_k} \cdot n = 0, \quad k \in I \text{ sample}, \qquad (9.9)$$

where I_{sample} is the set of indices for the sample components.

The validity of condition (9.9) stems from the fact that, as a rule, the samples are sufficiently far removed from the electrodes and do not participate in electrode reactions. Furthermore, in the case where the boundary Γ represents a gel plug, we may assume that the dimensions of the pores in the gel are much smaller than the dimensions of the molecules of the samples. We note that condition (9.9), together with (9.2), means that the total amount of sample in the mixture is conserved. It is easy to convince ourselves of this by integrating, for example, (II.4.14) over the entire volume of the mixture and using the Gauss–Ostrogradskii theorem:

$$\frac{\partial}{\partial t} \int_D \xi_k dv + \int_D \operatorname{div} i_{\xi_l} dv \equiv \frac{\partial}{\partial t} \int_D \xi_k dv + i_{\xi_k} \cdot n |_\Gamma + i_{\xi_k} \cdot n_0 |_{\Gamma_0} = 0,$$
$$\qquad (9.10)$$

$$\frac{\partial}{\partial t} \int_D \xi_k dv \equiv 0_0 \quad \xi_k \in I \text{ sample}.$$

In some concrete cases, conditions (9.9) may be replaced by conditions analogous to (9.6).

Although we have used the electric potential φ for classification of the boundaries, this does not at all mean that its value on the boundary Γ must be specified. A replacement for this condition may be the requirements that the tangential components of $\mathbf{E}$ go to zero at the boundary $\Gamma_0 + \Gamma$, and specification of the normal component of the current density at the boundary:

$$\Gamma : \mathbf{j} \cdot \mathbf{n} = J(t), \tag{9.11}$$

where $J(t)$ is the outward electric current density at the electrode.

In conclusion, let us recall that, in the case where the diffusion coefficients, mobilities, and other parameters do not depend on temperature, the thermal conductivity equation may be integrated independently. The boundary conditions for this equation, usually used in mathematical physics, will be laid out in each concrete case for solution of the given equation in the corresponding sections.

Chapter IV

ISOTACHOPHORESIS

An interesting and specific variant of electrophoresis occurs when the electrophoretic space is filled with a solution of electrolytes which yield one or several common ions upon dissociation. A typical example is a solution of several salts of the same acid or the same base. Under certain conditions in an applied electric field the solution is divided into pure zones of individual ions, and the common ion (the so-called counterion) is present everywhere. In this case, in the stabilized regime the boundaries between zones are sufficiently clear and move at a constant velocity, which is the basis for the name of this method; and the electrolytes to be separated line up in decreasing order of transport rates. In the general case, the order according to transport rates does not coincide with the order according to mobilities (see also [95]).

Underlying the current theory of isotachophoresis is the fundamental Kohlrausch's law (1887), which is a consequence of the law of conservation of charge and continuity of current. This law determines the connection between the concentrations of adjacent pure zones, and for a known current density makes it possible to determine the zone velocity. Kohlrausch's law is general in character and is the basis for the final steady-state theory. The physical meaning of this law is that the relative contribution of the common ion to the current is constant along the electrolyte. It is specifically on the basis of this law that transport numbers were introduced, which for each given ion express its fractional electrical conductivity relative to the electrical conductivity of the entire solution (for the measurement of transport numbers, see [20, 22]).

In a number of theoretical papers [46, 76, 95], Kohlrausch's law together with some of its immediate generalizations (in particular the case of several common ions) is used to describe steady states in different specific situations. An attempt at a more in-depth treatment based on the use of nonsteady-state partial differential equations is made in [85], where, for the variant of a two-component mixture which is reduced to the familiar Burgers equation, the development of a jump from an initially smooth concentration profile is studied numerically.

A complete theory of isotachophoresis should include a description of the process of establishing pure zones, the conditions for their generation, and the time it takes to form them, and also an investigation of the role of diffusion. These questions may be resolved within the framework of the nonsteady-state model to which this chapter is devoted. The model represents a system of nonlinear partial differential equations expressing the laws of conservation of mass and charge under conditions of electrical neutrality for the solution. In this system, the diffusion coefficients, mobilities, and molar charges are known functions of the unknown concentrations. When the chemistry of the processes occurring in solution is known, such functions are determined from the theory of local chemical equilibrium. The investigation is significantly facilitated by the fact that the effect of diffusion is localized in the vicinity of the boundary between two adjacent pure zones and is reduced to some blurring of this boundary. It is interesting to note that, within the pure zone, the electric forces lead to establishment of homogeneous concentrations (inhomogeneities lag behind the moving zone and, hitting its trailing boundary, are stopped by diffusion); the action of diffusion within the zone only promotes this process.

In the diffusion-free approximation, the model is reduced to a system of quasilinear hyperbolic equations, mainly analogous to the system of gas dynamics equations and corresponding to the class of systems which has been most completely studied in this theory: a system with a complete set of Riemann invariants. We also note that the indicated analogy [46] between isotachophoresis and elution and absorption chromatography goes deeper than just outward appearance. As study of the system shows, in the properties of both models we observe profound and far-reaching physicomathematical parallels. Thus, the role of the common ion completely corresponds to the role of the sorbent in absorption chromatography.

In the diffusion-free approximation the quasilinear hyperbolic system obtained allows us to answer the majority of questions posed above. Obviously, in the general case we need to use a computer and numerical methods; also, methods (in particular finite-difference methods) based on conservative difference schemes [40] have been well developed. For a whole series of cases important in practice (for example when the initial concentration distribution is piecewise-constant), the solution is obtained in analytic form, which is very convenient for application. On the basis of the solution of the so-called problem of the collapse of an arbitrary discontinuity, and also the problem of the interaction of progressive traveling waves, it proves to be possible to formulate a number of rules which allow us to describe the evolution of the solution from the initial state in the form of zone arrangement patterns and tables of the concentration distribution of the electrolytes in these zones. Each pattern, together with the corresponding table, gives a complete description of the process in a certain time segment. Transitions to the next pair of patterns and tables correspond to rearrangement of the zones caused by jump crossings. The transition times are calculated directly from the jump velocities indicated in the tables.

It is well known to experimenters that in order to separate electrolytes in isotachophoresis, the electrolytes must be located between the leader (which has the highest transport rate) and the terminator (which has the lowest transport rate); in this case, the amount of leader and terminator should be sufficiently high and the electric field should be oriented so that motion occurs from the terminator to the leader. In the literature there has been no analysis of phenomena arising when these conditions are violated. On the basis of the theory presented in this chapter, such phenomena are considered for a number of basic cases. The most characteristic phenomenon proves to be the *electrolytic mixing effect*, manifested as mixing zones described by progressive rarefaction waves (see Chapter V).

An interesting and rather unexpected consequence of the considered theory is an effect which we may call electrolytic memory. One of the Riemann invariants proves to be constant over time at each point of the electrolyte; i.e., the system remembers forever its shape specified at the initial moment. The existence of electrolytic memory considerably simplifies the computational aspect of the theory, and allows us to predict a number of important phenomena connected with isotachophoresis; and, moreover, it is possibly a basic feature distinguishing the isotachophoresis process from chromatography.

In this chapter the described approach is demonstrated for the simplest, most basic examples: separation of two partially mixed electrolytes (Section 5), separation of a two-component mixture between the terminator and the leader (Section 6), etc. From experiment it is well known that the motion of zones in isotachophoresis is accompanied by displacement of temperature jumps. The corresponding theory is given in Section 7.

1. Models of Isotachophoresis for Weak and Strong Electrolytes

In Chapter III we presented a model for isotachophoresis for arbitrary electrolytes [see (III.8.18)-(III.8.21)]. In this section we will consider two limiting cases of the application of this model, for strong and weak electrolytes. Rather unexpectedly, it has turned out that mathematically these cases are quite equivalent; i.e., they are described by the same equations. This is explained by the fact that, in both cases, albeit for different reasons, we may neglect diffusion processes.

Let us first consider the case of weak electrolytes. Let the dissociation constants of the mixture components have the form

$$K_i = K_* \alpha^2 \overline{K}_i, \tag{1.1}$$

where α is a small parameter. Let us restrict ourselves to consideration of the one-dimensional case, assuming that the electrophoretic chamber has definite length. Let us make the substitution of variables

$$a_1 = \alpha K_* \overline{a}_1, \ E = \frac{1}{\alpha} \overline{E}, \ \mu = \mu_0 \alpha, \ \mu_0 = O(1), \ \left(E = O\left(\frac{1}{\alpha}\right), a_1 = O(\alpha)\right), \tag{1.2}$$

$$\alpha \to 0.$$

We note that, in this model, the choice of the small parameter α is essentially connected with neglect of diffusion and with the choice of the magnitude of the electric field applied to the chamber. Since, by assumption, the electrolyte is weak, in order for an appreciable current to exist, we need to choose a higher field intensity. Obviously, in such a treatment the electrolytes are "weak" relative to the common ion (in this case, a positive ion); its concentration a_1 is much lower than the concentrations of the negative ions ξ_i.

Substituting (1.1) and (1.2) into (III.8.12)-(III.8.14), we obtain

$$\frac{\partial \xi_i}{\partial t} + \frac{\partial}{\partial x} \left\{ - \mu_0 \alpha \frac{\partial}{\partial x} \left(\frac{D_{2i+1} \alpha \overline{K}_i + D_{2i} \overline{a}_1}{\alpha \overline{K}_i + \overline{a}_1} \xi_i \right) \frac{D_{2i+1} \overline{K}_i \xi_i}{\alpha \overline{K}_i + \overline{a}_1} \overline{E} \right\} = 0,$$

$$(i = 1, \ldots, n) \tag{1.3}$$

$$\overline{a}_1 = \sum_{i=1}^{n} \frac{\overline{K}_i}{\alpha \overline{K}_i + \overline{a}_1} \xi_i, \tag{1.4}$$

$$j = - \mu_0 \alpha^2 \frac{\partial \overline{a}_1}{\partial x} + \mu_0 \frac{\partial}{\partial x} \alpha^2 \sum_{i=1}^{n} \frac{D_{2i+1} \overline{K}_i}{\alpha \overline{K}_i + \overline{a}_1} \xi_i + \left(\overline{a}_1 + \sum_{i=1}^{n} \frac{D_{2i+1} \overline{K}_i}{\alpha \overline{K}_i + \overline{a}_1} \xi_i \right) \overline{E}. \tag{1.5}$$

Discarding terms of order $O(\alpha)$ as $\alpha \to 0$, we derive the following system of equations (the bar over the quantities $\overline{K}_i, \overline{a}_1, \overline{E}$ is omitted for convenience):

$$\frac{\partial \xi_i}{\partial t} - \frac{\partial}{\partial x} \left(\frac{D_{2i+1} K_i \xi_i}{a_1} E \right) = 0 \quad (i = 1, \ldots, n), \tag{1.6}$$

$$a_1 = \sum_{i=1}^{n} \frac{K_i}{a_1} \xi_i, \quad j = \left(a_1 + \sum_{i=1}^{n} \frac{D_{2i+1}K_i}{a_1} \xi_i \right) E. \tag{1.7}$$

From (1.7), we easily obtain

$$\frac{E}{a_1} = \frac{j}{\sum\limits_{i=1}^{n} (1 + D_{2i+1}) K_i \xi_i}. \tag{1.8}$$

Substituting the obtained expression into (1.6), we have

$$\frac{\partial \xi_i}{\partial t} - \frac{\partial}{\partial x} \left\{ \frac{D_{2i+1}K_i \xi_i j}{\sum\limits_{i=1}^{n} (1 + D_{2i+1}) K_i \xi_i} \right\} = 0 \quad (i = 1, \dots, n). \tag{1.9}$$

We make the substitutions

$$u_i = (1 + D_{2i+1}) K_i \xi_i, \quad \xi_i = \frac{u_i}{(1 + D_{2i+1}) K_i}, \quad \alpha_i = -jK_iD_{2i+1}. \tag{1.10}$$

Then, finally, system (1.9) is rewritten in the form [compare with (III.8.18)-(III.8.21)]

$$\frac{\partial u_i}{\partial t} + \frac{\partial}{\partial x} \left(\frac{\alpha_i u_i}{\sum\limits_{k=1}^{n} u_k} \right) = 0 \quad (i = 1, \dots, n). \tag{1.11}$$

The system of equations obtained describes the isotachophoresis process for weak electrolytes. After determining u_i from (1.11), the values of the concentration of the common ion a_1 and the electric field intensity are determined from (1.7) using (1.10).

Let us now turn to the case of strong electrolytes, where the dissociation constants are sufficiently large: complete decomposition of all mixture components into ions occurs. Let the following relations be satisfied:

$$K_i = \frac{\overline{K}_i}{\alpha}, \quad \overline{K}_i = O(1), \quad E = O(1), \quad \mu = \mu_0 \alpha, \quad \mu_0 = O(1), \quad a_1 = O(1), \tag{1.12}$$

$$\xi_i = O(1),$$

where α is a small parameter determined essentially by diffusion in the system. Substituting (1.12) into (III.8.18)-(III.8.21), we obtain

$$\frac{\partial \xi_i}{\partial t} + \frac{\partial}{\partial x} \left\{ -\alpha \mu_0 \frac{\partial}{\partial x} \left(\frac{D_{2i+1}\overline{K}_i + D_{2i}\alpha a_1}{\overline{K}_i + \alpha a_1} \xi_i \right) - \frac{D_{2i+1}\overline{K}_i \xi_i}{\overline{K}_i + \alpha a_1} E \right\}$$
$$= 0 \quad (i = 1, \dots, n), \tag{1.13}$$

$$a_1 = \sum_{i=1}^{n} \frac{\overline{K}_i}{\overline{K}_i + \alpha a_1} \, \xi_i, \tag{1.14}$$

$$\hat{\jmath} = -\alpha\mu_0 \frac{\partial a_i}{\partial x} + \alpha\mu_0 \frac{\partial}{\partial x} \sum_{i=1}^{n} \frac{D_{2i+1}\overline{K}_i}{\overline{K}_i + a_1} \, \xi_i + \left(a_1 + \sum_{i=1}^{n} \frac{D_{2i+1}\overline{K}_i}{\overline{K}_i + \alpha a_1} \, \xi_i \right) E. \tag{1.15}$$

Discarding terms of order $O(\alpha)$, $\alpha \to 0$, and making substitutions analogous to those in (1.10), we obtain

$$\frac{\partial v_i}{\partial t} + \frac{\partial}{\partial x} \left(\frac{\beta_i v_i}{\sum\limits_{k=1}^{n} v_k} \right) = 0 \quad (i = 1, \ldots, n), \tag{1.16}$$

$$\beta_i = - j D_{2i+2}, \quad v_i = (1 + D_{2i+1}) \, \xi_i. \tag{1.17}$$

Thus, for strong electrolytes the isotachophoresis process is described by equations equivalent to (1.11).

Let us recall that the current density j in the general case may depend on time; consequently, the coefficients α_i and β_i in Eqs. (1.11) and (1.17) are also functions of time. However, as a rule the isotachophoresis process is carried out at constant current. Therefore, in the following development we will restrict ourselves to consideration of specifically this case, and we will consider the coefficients α_i (or β_i) to be constant. We note that if the function $j(t)$ does not change sign, then by substitution of variables

$$\bar{\alpha}_i \to \alpha_i \frac{1}{j(t)}, \quad \tau = \tau(t) = \int_0^t j(s) \, ds, \tag{1.18}$$

the system is reduced to the previous form and $\bar{\alpha}_i = \text{const}$ $(i = 1, \ldots, n)$.

2. Riemann Invariants of the System of Quasilinear Equations

This section is devoted to investigation of the system of equations (1.11), corresponding to the class of quasilinear systems of the hyperbolic type [40]. There are many difficulties in the general theory of such systems. Fortunately, however, system (1.11) belongs to the class of systems which has been investigated in depth by Rozhdestvenskii [40, 41]. In the following development, we will mainly follow the excellent work by Kuznetsov [25], who studied the quasilinear system of equations describing the processes of sorption elution chromatography. In the simplest case of Langmuir sorption, the latter has the form

$$\frac{\partial}{\partial t}\left(v_i + \frac{\Gamma_i v_i}{p}\right) + V\frac{\partial v_i}{\partial x} = 0, \quad p = 1 + \sum_{s=1}^{n} v_s,$$

where v_i is the concentration of the ith component, $\Gamma_i =$ const are the Henry coefficients, and V is the flow velocity.

The similarity of this system to (1.11) is rather obvious; below we will consider the connection between them in more detail. Many authors (see, for example, [46]) have pointed out the analogy between isotachophoresis and elution chromatography. The material presented in this section allows us to clarify this analogy. In particular, it turns out that the role of the common ion of the electrolytes in isotachophoresis is analogous to the role of the sorbent in chromatography.

Thus, we will consider the quasilinear system of equations (1.11) for unknown concentrations u_i with transport rates α_i [see (1.10)]:

$$\frac{\partial u_i}{\partial t} + \frac{\partial}{\partial x}\frac{\alpha_i u_i}{s} = 0 \quad (i = 1, \ldots n), \tag{2.1}$$

where

$$s = \sum_{k=1}^{n} u_k. \tag{2.2}$$

Let us introduce the symbols

$$\varphi_i = \varphi_i(u_1, \ldots, u_n) = \frac{\alpha_i u_i}{s}, \quad A_{ij} = \frac{\partial \varphi_i}{\partial u_j} = \alpha_i \frac{\delta_{ij}}{s} - \alpha_i u_i \frac{1}{s^2}. \tag{2.3}$$

Then (2.1) takes on the form

$$\frac{\partial u_i}{\partial t} + \sum_{j=1}^{n} A_{ij}\frac{\partial u_j}{\partial x} = 0 \quad (i = 1, \ldots, n). \tag{2.4}$$

Let us show that (2.4) may be rewritten in Riemann invariants

$$\frac{\partial R_i}{\partial t} + \xi^{(i)}\frac{\partial R_i}{\partial x} = 0 \quad (i = 1, \ldots, n), \tag{2.5}$$

where

$$R_i = R_i(u_1, \ldots, u_n), \quad \xi^{(i)} = \xi^{(i)}(u_1, \ldots, u_n). \tag{2.6}$$

Let us determine the right-hand $[r_i^{(k)} = (r_1^k, \ldots, r_n^k)]$ and the left-hand $[l_i^{(k)} = (l_1^k, \ldots, l_n^k)], k = 1, 2, \ldots, n$, eigenvectors of the matrix $(A_{ji})_{ij=1}^n = 1$ corresponding to the eigenvalues $\xi^{(k)}$. In the following development, we will omit the superscripts k for convenience when possible. Thus, we have

$$\sum_{j=1}^n A_{ij} r_j \equiv \xi r_i, \quad \sum_{i=1}^n A_{ij} l_i \equiv \xi l_j. \tag{2.7}$$

Substituting (2.3) into (2.7), we obtain for the right-hand eigenvector

$$\frac{\alpha_i r_i}{s} - \sum_{j=1}^n \left(\frac{\alpha_i u_i}{s^2} \right) r_j \equiv \xi r_i \quad (i = 1, \ldots, n). \tag{2.8}$$

From this

$$r_i = \frac{\alpha_i u_i}{\alpha_i s - s^2 \xi} \sum_{j=1}^n r_j. \tag{2.9}$$

Therefore, normalizing the right-hand eigenvector r with the condition $\sum_{j=1}^n r_j = 1$, we obtain

$$r_i = \frac{\alpha_i u_i}{\alpha_i s - \lambda}, \tag{2.10}$$

where, for convenience, we introduce the symbol

$$\lambda \equiv \xi s^2. \tag{2.11}$$

Analogously, for the left-hand eigenvector we derive

$$l_i = \frac{1}{\alpha_i s - \lambda}. \tag{2.12}$$

We note for later use the equality

$$\sum_{i=1}^n r_i = 1. \tag{2.13}$$

Substituting (2.10) into this expression, we obtain an equation for determining the eigenvalue of the matrix A as a function of $u_1, \ldots, u_n$:

$$\sum_{i=1}^{n} \frac{\alpha_i u_i}{\alpha_i s - \lambda} = 1, \quad \lambda = \xi s^2. \tag{2.14}$$

By virtue of the equality (2.2), $\xi = \xi^{(1)} = 0$ is a root of Eq. (2.14), i.e., an eigenvalue of the matrix A.

Let the quantity α_i be positive and numbered in increasing order

$$0 < \alpha_1 < \alpha_2 < \alpha_3 < \ldots < \alpha_n. \tag{2.15}$$

Then for positive values of $u_1, \ldots, u_n$, the function

$$F(\xi) \equiv \sum_{i=1}^{n} \frac{\alpha_i u_i}{\alpha_i s - \xi s^2} \tag{2.16}$$

has the form shown in Fig. 5 (compare with [25]), where

$$\xi_i^* = \frac{\alpha_i}{s}. \tag{2.17}$$

Thus, Eq. (2.14) has n real, different roots $\xi^{(i)}$; the matrix A_{ij} has n real, simple eigenvalues located in the interval $[0, \xi_n]$ and satisfying the inequalities

$$\xi_1^* < \xi^{(2)} < \xi_2^*, \quad \xi_{i-1}^* < \xi^{(i)} < \xi_i^*. \tag{2.18}$$

Consequently, system (2.1) for $\alpha_i u_i \neq 0$ and positive $u_1, \ldots, u_n$ is hyperbolic in the narrow sense [40].

Multiplying system (2.4) by the left-hand eigenvector, we obtain

$$\sum_{i=1}^{n} l_i^{(k)} \left(\frac{\partial u_i}{\partial t} + \xi^{(k)} \frac{\partial u_i}{\partial x} \right) = 0 \quad (k = 1, \ldots, n). \tag{2.19}$$

If the following relationship is satisfied:

$$\frac{\partial R_k}{\partial u_i} = \mu_k l_i^{(k)} \quad (k = 1, \ldots, n), \tag{2.20}$$

then (2.19) takes on the form (2.5), i.e., $R_k(u_1, \ldots, u_n)$ are Riemann invariants. Condition (2.20) means that the differential forms

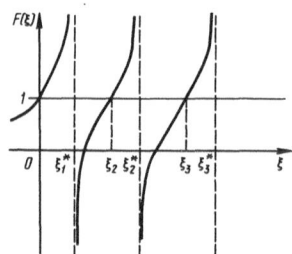

Fig. 5. Form of the function $F(\xi)$.

$$\omega_k \equiv \sum_{i=1}^{n} l_i^{(k)} du_i \qquad (2.21)$$

are integrable with the factor μ_k.

Let us show that as the Riemann invariants R_k $(k = 2, \ldots, n)$ we may choose the quantities

$$R_k = R_k(u_1, \ldots, u_n) = s\xi^{(k)} \quad (k = 2, \ldots, n). \qquad (2.22)$$

Differentiating (2.14) with respect to u_j, we obtain

$$0 = \frac{\alpha_j}{\alpha_j s - \lambda} - \sum_{i=1}^{n} \frac{\alpha_i^2 u_i}{(\alpha_i s - \lambda)^2} + \frac{\partial \lambda}{\partial u_j} \sum_{i=1}^{n} \frac{\alpha_i u_i}{(\alpha_i s - \lambda)^2} \equiv \frac{\alpha_j}{\alpha_j s - \lambda}$$

$$- \sum_{i=1}^{n} \alpha_i r_i l_i + \frac{\partial \lambda}{\partial u_j} \sum_{i=1}^{n} r_i l_i. \qquad (2.23)$$

From this we have

$$\frac{\partial \lambda}{\partial u_j} = \frac{\displaystyle\sum_{i=1}^{n} \alpha_i r_i l_i - \frac{\alpha_j}{\alpha_j s - \lambda}}{\displaystyle\sum_{i=1}^{n} r_i l_i}. \qquad (2.24)$$

From (2.22) and (2.11), we obtain

$$\frac{\partial R}{\partial u_j} \equiv \frac{\partial}{\partial u_j}\left(\frac{\lambda}{s}\right) \equiv \frac{1}{s}\left(\frac{\partial \lambda}{\partial u_j} - \frac{\lambda}{s}\right). \qquad (2.25)$$

Substituting here (2.24) and taking into account (2.13), we derive

$$\frac{\partial R}{\partial u_j} = \frac{\sum_{i=1}^{n}\left(\alpha_i r_i l_i - \frac{\lambda}{s} r_i l_i\right) - \frac{\alpha_j}{\alpha_j s - \lambda}}{s \sum_{i=1}^{n} r_i l_i} \equiv \frac{\frac{1}{s} - \frac{\alpha_j}{\alpha_j s - \lambda}}{s \sum_{i=1}^{n} r_i l_i} \equiv \frac{-\lambda l_j}{s^2 \sum_{i=1}^{n} r_i l_i}$$

$$(2.26)$$

Thus, relationship (2.20) is satisfied, and

$$\mu_k \equiv \frac{-\xi^{(k)}}{\sum_{i=1}^{n} r_i^{(k)} l_i^{(k)}} \qquad (k = 2, \ldots, n). \tag{2.27}$$

The Riemann invariant R_1 corresponding to the eigenvalue $\xi^{(1)} = 0$ is easy to obtain, dividing (2.1) by α_i and summing over all i

$$\frac{\partial}{\partial t} \sum_{i=1}^{n} \frac{u_i}{\alpha_i} = 0, \quad R_1 \equiv \sum_{i=1}^{n} \frac{u_i}{\alpha_i}. \tag{2.28}$$

Finally, system (2.1) in Riemann invariants is rewritten in the form

$$\frac{\partial R_1}{\partial t} = 0, \tag{2.29}$$

$$\frac{\partial R_k}{\partial t} + \xi^{(k)} \frac{\partial R_k}{\partial x} = 0, \quad R_k = s\xi^{(k)} \quad (k = 2, \ldots, n). \tag{2.30}$$

In the following development, we will be interested in the generalized (discontinuous) solutions to (2.1). Therefore, let us write the conditions on the lines of discontinuity. We will consider system (2.1) as a consequence of the integral conservation laws

$$\oint_C \left(u_i dx - \frac{\alpha_i u_i}{s} dt\right) = 0. \tag{2.31}$$

Here, C is a piecewise-smooth contour in the region $D \subset R_1^2$, where the solution $u_1(x, t), \ldots, u_n(x, t)$ is defined. Let $x = x(t)$ be the equation of the line of discontinuity. Let us use the symbols

$$D = x'(t), \quad u^-(t) = u(x(t) - 0, t), \quad u^+(t) = u(x(t) + 0, t),$$

$$[u] = u^+(t) - u^-(t). \tag{2.32}$$

The superscripts plus and minus correspond to the solutions, respectively, to the right and to the left of the jump. Then, on the lines of discontinuity, the so-called Hugoniot conditions are satisfied at the jumps:

$$D \cdot [u_i] = \left[\frac{\alpha_i u_i}{s} \right] \quad (i = 1, \ldots, n). \tag{2.33}$$

Dividing this expression by α_i and summing over all i, we derive for R_1 the equality

$$D[R_1] = 0. \tag{2.34}$$

From the equality (2.34), we have the following important consequences:

1. On moving jumps, the invariant $R_1(u_1, \ldots, u_n)$ does not have a discontinuity for any generalized solution.

2. The jumps of the invariant R_1 are fixed and, consequently, are defined by the initial data. At the fixed jumps, all the quantities u_i/s $(i = 1, \ldots, n)$ are continuous.

Later we consider some solutions to system (2.1), when all the concentrations are piecewise-constant. For these cases, the solutions are obtained in explicit and easily discernable form.

3. Motion of Two Zones with Arbitrary Concentration, Separated at the Initial Instant of Time

Let us consider the case where an infinite electrophoretic column is filled with two electrolytes with constant concentrations. At the initial instant of time, let these electrolytes be separated at the point $x = 0$ (Fig. 6), where $\xi_i^{(0)}$ are the initial concentrations of the electrolytes.

The system of equations describing the isotachophoresis process in Riemann invariants has the form ($n = 2$) (see also [85])

$$\frac{\partial R_1}{\partial t} = 0, \tag{3.1}$$

$$\frac{\partial R_2}{\partial t} + \xi^{(2)} \frac{\partial R_2}{\partial x} = 0. \tag{3.2}$$

In this case, the dependence of the eigenvalue $\xi^{(2)}$ on u_1, u_2 is easily determined using (2.14), taking on the form

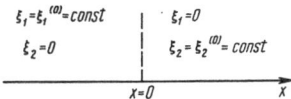

Fig. 6. Initial concentration distribution in the column.

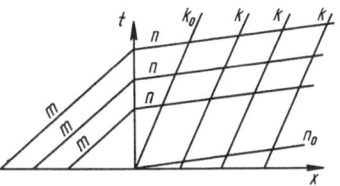

Fig. 7. Form of the characteristics corresponding to Eq. (3.11).

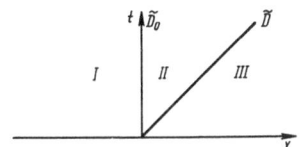

Fig. 8. Lines of discontinuity on the (x, t) plane.

$$\frac{\alpha_1 u_1}{\alpha_1 s - \xi s^2} + \frac{\alpha_2 u_2}{\alpha_{,s} - \xi s^2} = 1. \tag{3.3}$$

From this we derive

$$\xi^{(2)} = \frac{\alpha_1 u_2 + \alpha_2 u_1}{(u_1 + u_2)^2}, \quad \xi^{(1)} = 0. \tag{3.4}$$

For the Riemann invariants, from (2.28) and (2.30) we have the expressions

$$R_1 = \frac{u_1}{\alpha_1} + \frac{u_2}{\alpha_2}, \quad R_2 = \frac{\alpha_1 u_2 + \alpha_2 u_1}{u_1 + u_2}. \tag{3.5}$$

For concreteness, we will consider weak electrolytes. Let us introduce the symbols [see (1.10)]

$$v_1 = (1 + D_3) K_1 \xi_1^{(0)} \neq 0, \quad v_2 = (1 + D_5) K_2 \xi_2^{(0)} \neq 0, \tag{3.6}$$

$$\alpha_1 \equiv - j K_1 D_3 > 0, \tag{3.7}$$

$$\alpha_2 = - j K_2 D_5 > 0, \quad j < 0.$$

Then the initial conditions have the form

$$u_1 = \begin{cases} v_1, & x < 0, \\ 0, & x > 0 \end{cases} \quad u_2 = \begin{cases} 0, & x < 0 \\ v_2, & x > 0. \end{cases} \tag{3.8}$$

For system (3.1), (3.2), the initial conditions will be [see (3.5)]

$$R_1 = \begin{cases} \dfrac{v_1}{\alpha_1}, & x < 0 \\ \dfrac{v_2}{\alpha_2}, & x > 0, \end{cases} \quad R_2 = \begin{cases} \alpha_2, & x < 0 \\ \alpha_1, & x > 0. \end{cases} \tag{3.9}$$

Let us establish the discontinuous piecewise-constant solution to system (3.1), (3.2) with conditions (3.9). From (3.1), it is obvious that

$$R_1 (x, \ t) = R_1 (x) = \begin{cases} v_1/\alpha_1, & x < 0 \\ v_2/\alpha_2, & x > 0. \end{cases} \tag{3.10}$$

The equation of the characteristics on the $(x, \ t)$ plane for (3.2) has the form

$$\frac{dx}{dt} = \xi^{(2)} \equiv \frac{R_2^2}{\alpha_1 \alpha_2 R_1 (x)} . \tag{3.11}$$

We rewrite (3.2) in the form

$$\frac{dR_2}{dt} = 0, \quad \frac{d(\ )}{dt} = \frac{\partial (\ )}{\partial t} + \frac{dx}{dt} \frac{\partial (\ )}{\partial x} . \tag{3.12}$$

From this,

$$R_2 = R_2 (x, \ t) = R_2 (a (x, \ t)) = R_2 (a), \tag{3.13}$$

where $a \equiv a(x, \ t)$ is the point of intersection of the characteristic passing through the point $(x, \ t)$ and the line $(x, 0)$ on the $(x, \ t)$ plane.

Integrating (3.11) and taking into account (3.13), we obtain

$$\alpha_1 \alpha_2 \int_a^x R_1 (y) \, dy = R_2^2 (a) t. \tag{3.14}$$

From this we determine the characteristics (Fig. 7)

$$x = \frac{\alpha_2}{v_1} t + a, \ x < 0, \ a < 0 \ \text{(lines } m\text{)}; \tag{3.15}$$

$$x = \frac{\alpha_2}{\alpha_1 v_2} t + \frac{\alpha_2 v_1}{\alpha_1 v_2} a, \ x > 0, \ a < 0 \ \text{(lines } n\text{)}; \tag{3.16}$$

$$x = \frac{\alpha_1}{v_2} t + a, \ x > 0, \ a > 0 \ \text{(lines } k)_{\downarrow} \tag{3.17}$$

On the characteristic k_0, the velocity du/dt of the "particles" is

$$\frac{dx}{dt} = \xi^+, \ \xi^+ \equiv \frac{R_2^2(+0)}{\alpha_1 \alpha_2 R_1(+0)} = \frac{\alpha_1}{v_2}, \tag{3.18}$$

and on the characteristic n_0

$$\frac{dx}{dt} = \xi^-,$$

$$\xi^- \equiv \frac{R_2^2(-0)}{\alpha_1 \alpha_2 R_1(+0)} = \frac{\alpha_2^2}{\alpha_1 v_2}. \tag{3.19}$$

Let the following relationship be satisfied (the condition for existence of a moving discontinuity, following from the jump stability condition):

$$\xi^- > \xi^+. \tag{3.20}$$

For this, it is sufficient to satisfy the condition

$$\alpha_2 > \alpha_1 \tag{3.21}$$

or [see (3.7)]

$$K_1 D_3 < K_2 D_5. \tag{3.22}$$

The latter means that we find the "faster" electrolyte (the one having the higher transport rate) in front along the path of motion in an electric field.

Let us introduce the lines of discontinuity $\check{D}_0$, $\check{D}$ and the symbols (Fig. 8)

$$\left. \begin{array}{l} u_1 = \eta_1, \ u_2 = \zeta_1 \ \text{(in region I)}, \\ u_1 = \eta_2; \ u_2 = \zeta_2 \ \text{(in region II)}, \\ u_1 = \eta_3, \ u_3 = \zeta_3 \ \text{(in region III)}, \end{array} \right\} \tag{3.23}$$

Let D_0 and D be the velocities of the discontinuities. On the line $\check{D}_0$, the quantity R_1 undergoes a discontinuity. Then from the Hugoniot conditions (2.34) it follows that the velocity of the discontinuity is equal to zero:

$$[R_1] \neq 0 \ \text{on the line } \bar{D}_0 \, (D_0 = 0). \tag{3.24}$$

Taking into account the fact that

$$\eta_1 = v_1, \ \zeta_1 = 0; \ \eta_3 = 0, \ \zeta_3 = v_2, \tag{3.25}$$

we rewrite the Hugoniot conditions

$$-\frac{2}{\eta_2+\zeta_2}=0,\; \frac{\eta_2}{\eta_2+\zeta_2}-1=0,\; D\eta_2=\frac{\alpha_1}{\eta_2+\zeta_2}\,\eta_2,\; \frac{\eta_2}{\alpha_1}+\frac{\zeta_2}{\alpha_2}=\frac{v_2}{\alpha_2}.$$

$$(3.26)$$

From this,

$$\zeta_2=0,\quad \eta_2=\frac{\alpha_1}{\alpha_2}\,v_2,\qquad D=\frac{\alpha_2}{v_2}.\qquad (3.27)$$

Thus, the continuous piecewise-constant (generalized) solution has the form

$$u_1=\begin{cases} v_1, & x<0, & t>0,\\[2mm] \dfrac{\alpha_1}{\alpha_2}\,v_2, & 0<x<\dfrac{\alpha_2}{v_2}\,t, & t>0,\\[2mm] 0, & x>\dfrac{\alpha_2}{v_2}\,t, & t>0; \end{cases} \qquad (3.28)$$

$$u_2=\begin{cases} 0, & x<0, & t>0,\\[2mm] 0, & 0<x<\dfrac{\alpha_2}{v_2}\,t, & t>0,\\[2mm] v_2, & x>\dfrac{\alpha_2}{v_2}\,t, & t>0. \end{cases} \qquad (3.29)$$

Using (1.10) and (3.6), it is not difficult to go to the concentrations (Fig. 9)

$$\xi_1(x,\;t)=\begin{cases} \xi_1^{(0)}, & x<0.\\[2mm] \dfrac{D_3(1+D_5)}{D_5(1+D_3)}\cdot\xi_2^{(0)}, & 0<x<-\dfrac{jD_5}{(1+D_5)\,\xi_2^{(0)}}\,t;\\[3mm] 0, & x>-\dfrac{jD_5}{(1+D_5)\,\xi_2^{(0)}}\,t, \end{cases} \qquad (3.30)$$

$$\xi_2(x,\;t)=\begin{cases} 0, & x<0,\\[2mm] 0, & 0<x<-\dfrac{jD_5}{(1+D_5)\,\xi_2^{(0)}}\,t,\\[3mm] \xi_2^{(0)}, & x>-\dfrac{jD_5}{(1+D_5)\,\xi_2^{(0)}}\,t. \end{cases} \qquad (3.31)$$

The interface between the two electrolytes moves with the velocity $-j[D_5/(1+D_5)\xi_2^0]$, completely determined by the leading electrolyte. The ratio of the concentrations at the interface satisfies the familiar Kohlrausch condition (see, for example, [46, 95]), which in dimensional variables has the form

$$\frac{\xi_1}{\xi_2^{(0)}}=\frac{D_3(D_1+D_5)}{D_5(D_1+D_3)}, \qquad (3.32)$$

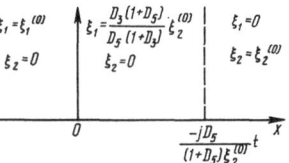

Fig. 9. Final concentration distribution in the column.

where D_1 is the diffusion coefficient of the common positive ion. We recall that for singly charged electrolytes when the Einstein relations are satisfied, the diffusion coefficients are proportional to the mobilities.

The concentration distributions of the common ion and the electric field intensities are as follows:

$$a_1(x,\ t) = \begin{cases} \sqrt{K_1\xi_1^{(0)}}\,, & x<0,\ t>0, \\[2mm] \sqrt{K_1\dfrac{D_3(1+D_5)}{D_5(1+D_3)}\,\xi_2^{(0)}}\,, & 0<x<-j\dfrac{D_5 t}{(1+D_5)\,\xi_2^{(0)}}\,, & t>0, \\[2mm] \sqrt{K_2\xi_2^{(0)}}\,, & -j\dfrac{D_5 t}{(1+D_5)\,\xi_2^{(0)}}<x, & t>0; \end{cases}$$

$$(3.33)$$

$$E(x,\ t) = \begin{cases} \dfrac{j}{(1+D_3)\sqrt{K_1\xi_1^{(0)}}}\,, & x<0, & t>0, \\[3mm] \dfrac{j}{(1+D_3)\sqrt{K_1\dfrac{D_3(1+D_5)}{D_5(1+D_3)}\,\xi_2^{(0)}}}\,, & 0<x<-j\dfrac{D_5}{(1+D_5)\xi_2^{(0)}}t, & t>0, \\[3mm] \dfrac{j}{(1+D_5)\sqrt{K_2\xi_2^{(0)}}}\,, & -j\dfrac{D_5}{(1+D_5)\,\xi_2^{(0)}}t<x, & t>0. \end{cases}$$

$$(3.34)$$

From what has been presented, it follows that if at the initial instant of time the two electrolytes are completely separated and the electrolyte which has the higher transport rate is located "downstream," the process occurs as follows: the fast electrolyte moves against the field (we recall that the ions are negatively charged), retaining the constant initial concentration. As far as the slow electrolyte is concerned, its concentration remains constant and equal to the initial concentration only in the region which it occupies initially. At the point of the column where the interface between the electrolytes is located at the initial instant of time, a constant concentration jump arises for the slow electrolyte; and upon going through this jump, the slow electrolyte takes on a concentration which satisfies Kohlrausch's law at the moving jump: $u_1 = (\alpha_1/\alpha_2)v_2$. The moving jump travels with a velocity α_2/ν_2. Thus, both the velocity of the jump and the concentration of the slow electrolyte are completely determined by the parameters of the leader. It is especially interesting that the concentration of the slow electrolyte to the right of the initial fixed jump does not depend on its initial concentration; and, consequently, it may be brought as close as desired to the Kohlrausch value $(\alpha_1/\alpha_2)\nu_2$.

From the preceding, obviously the unrealistic conclusion also follows that this concentration occurs instantaneously upon passing through the fixed jump. This conclusion is obtained because the chemical reactions were assumed to be infinitely fast. Consequently, the concentration time is determined by the rate constants of the chemical reactions, and in order to determine this time we need to considerably improve the model. In this case, we should also take into account the effect of diffusion.

In the case where the slower electrolyte is located "downstream," problems (3.1), (3.2), (3.9) also may be solved by the methods in [25, 41]. In this case, the solution proves to be more complicated: the property of piecewise-constancy is lost (a piecewise-constant solution exists, but for this solution the familiar jump stability condition is not satisfied), and the so-called rarefaction waves arise. We will call this phenomenon "electrolytic mixing": the electric field preferentially mixes the electrolytes rather than separating them.

4. Case of Any Number of Zones of Pure Electrolytes

In this section, the result of the preceding section is generalized for the case where, at the initial instant of time, the column is filled with electrolytes having constant concentrations which are completely separated. However, in contrast to the preceding section, we will not trace out the trajectory of the liquid particles in detail, but rather we will consider only motion of the zones, i.e., the motion of the jumps delimiting the zones. In this case, the discussion is considerably simplified, and the results may be formulated in the general case of n components (see rules 1-8 at the end of this section). If for any reason we need to know the law of motion for the individual liquid particles making up the zones (for example isotopically labeled particles), we must turn to the equations of the characteristics. In the case of n separated zones, the constructions are quite analogous to those presented in Section 3 for $n = 2$. The corresponding initial conditions for the solutions to system (2.1) have the form (Fig. 10)

$$u_i|_{t=0} \begin{cases} v_i, & a_{i-1} < x < a_i; \\ 0, & x \bar\in (a_{i-1}, \, a_i) \end{cases} \quad i = 1, 2, \ldots n. \qquad (4.1)$$

Fig. 10. Initial arrangement of zones when $n = 4$. Explanations in text.

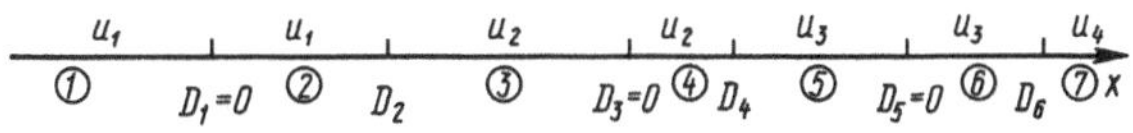

Fig. 11. Zones developing from the initial state. Above each zone is indicated that concentration in the zone which is different from zero.

TABLE 4. Concentration of Substances in the Zones and Jump Velocities

Index of substance (i)	Zone number (k)						
	1	2	3	4	5	6	7
1	v_1	$\alpha_1 \dfrac{v_2}{\alpha_2}$	0	0	0	0	0
2	0	0	v_2	$\alpha_2 \dfrac{v_3}{\alpha_3}$	0	0	0
3	0	0	0	0	v_3	$\dfrac{\alpha_3}{\alpha_4} v_4$	0
4	0	0	0	0	0	0	v_4
Jump velocity for leading edge of kth zone (D_k)	0	$\dfrac{\alpha_2}{v_2}$	0	$\dfrac{\alpha_3}{v_3}$	0	$\dfrac{\alpha_4}{v_4}$	—

Here, $v_i > 0$, $i = 1, 2, \ldots, n$, and we assume that $a_0 = -\infty$, $a_n = +\infty$. The column is assumed to be infinite; i.e., we do not consider end effects. Since the system under consideration is hyperbolic, disturbances are propagated with finite velocity, and, consequently, the solution obtained remains valid even for each given point of a finite column as long as disturbances do not arrive at that point from the ends. In general, the value of the solution $u(x, t)$ at the given point (x, t) depends only on the initial data from a finite segment (the region of influence). We note that the effect of diffusion is propagated instantaneously, but rapidly decreases as we move away from the source of the disturbance. As in the case where diffusion is taken into account, the assumption that the column is infinitely long is valid if the column is long enough.

Later we will establish a piecewise-constant generalized solution to (2.1) with initial condition (4.1). We immediately note that the jump stability condition leads to the requirement that the electrolytes "downstream" (the direction of "flow" is the positive direction of the x axis) be ordered according to transport rates so that $\alpha_1 < \alpha_2 < \ldots < \alpha_n$.

This condition is then assumed to be satisfied. Violation of this condition entails the appearance of rarefaction waves – the solution stops being piecewise-constant (see Chapter V, Section 2). We should also note that solutions with jumps describe irreversible processes; in shock waves it is already impossible to neglect diffusion (and if macroscopic motion occurs, then it is also impossible to neglect viscosity and thermal conductivity). It is precisely allowance for diffusion close to the jumps that leads to the jump stability conditions. Thus, the solutions presented below do not allow for time reversal.

In the following discussion, for concreteness let us assume $n = 4$; the results are transferred to any value of n in an obvious manner. In the following we will call a "zone" the segment of column between two successive jumps. We will call a zone "pure" if it contains only one electrolyte, and we will call a zone "mixed" if it contains several electrolytes. We will call a pure zone containing only the ith component the "u_i zone." The problem involves describing all the zones and their motion along the column. In Fig. 11, we illustrate the zones developing from the initial state (in Fig. 10, immediately after the onset of the process). Altogether there are seven zones (they are labeled by the numbers in circles), while in the initial state there were only four zones. We have designated the jump velocities as $D_1, D_2, ..., D_6$. In the following, we will also designate the jumps themselves as $D_1, D_2, ..., D_6$. This should hardly cause any confusion.

In order to establish the solution, first of all we note that the Riemann invariant R_1 does not depend on time, so that for all $t > 0$

$$R_1 = \frac{u_1}{\alpha_1} + \frac{u_2}{\alpha_2} + \frac{u_3}{\alpha_3} + \frac{u_4}{\alpha_4} = \frac{v_i}{\alpha_i}, \quad \text{if} \quad a_{i-1} < x < a_i. \quad (4.2)$$

Thus, for all $t > 0$, fixed jumps continue to exist at the points $x = a_i$ ($i = 1, 2, 3, 4$), $D_1 = D_3 = D_5 = 0$; all the rest of the jumps are moving jumps.

Let us now show that zone 2 is pure and contains only the first component. Let us further agree to call the concentrations in the kth zone $u_1^k, u_2^k, u_3^k, u_4^k$, and let us agree to call their sum s_k: $s_k = u_1^k + u_2^k + u_3^k + u_4^k$. We need to show that $u_2^2, u_3^2 = u_4^2 = 0$. Going back to the jump condition (2.33) we see that in the case where the jump is fixed and $D = 0$, all the quantities in u_i/s do not experience discontinuity. Therefore, if any component is missing on one side of the discontinuity, then it does not exist on the other side; more formally, we have the equality

$$\frac{u_i^1}{s_1} = \frac{u_i^2}{s_2} \quad (i = 1, 2, 3, 4). \quad (4.3)$$

Since $u_2^1 = u_3^1 = u_4^1 = 0$, from (4.3) it follows that $u_2^2 = u_3^2 = u_4^2 = 0$; condition (4.3) for $i = 1$ is converted to the trivial equality $1 = 1$.

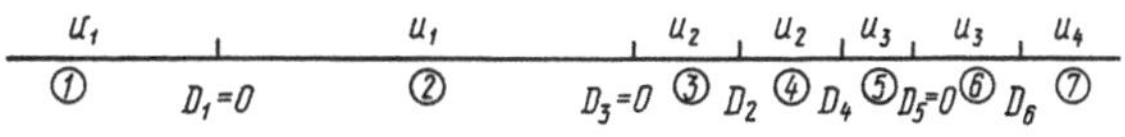

Fig. 12. Zone arrangement in the time interval $t_{23} < t < t_{45}$; t_{km} is the instant that jumps D_k and D_m cross.

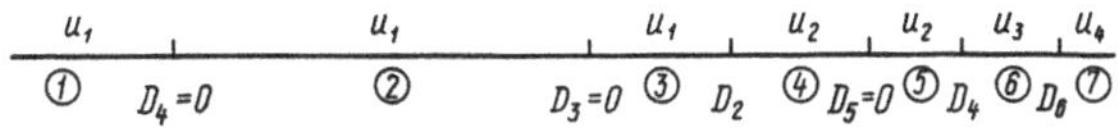

Fig. 13. Arrangement of zones in the time interval $t_{45} < t < t_{25}$.

Zone 2 is separated from the third fixed jump propagating with velocity D_2. From the hyperbolicity of the considered system (2.1), the positivity of the velocities of all the quasiparticles, and the fact that any set of constants is a solution, it follows that the solution at point x on the segment between D_1 and D_3 may experience a change only after the jump D_2 passes through the point. Thus, zone 3 is pure and $u_2 = v_2$ in that zone. We determine the velocity v_2 from the jump conditions

$$D_2 (u_i^2 - u_i^3) = \alpha_i \left(\frac{u_i^2}{s_2} - \frac{u_i^3}{s_3} \right), \quad i = 1, 2, 3, 4. \qquad (4.4)$$

When $i = 3, 4$, these equalities are satisfied identically; and setting $i = 1$ and taking into account the fact that $u_1{}^2 \neq 0$, $u_1{}^3 = 0$, we find that

$$D_2 u_1^2 = \alpha_1 \frac{u_1^2}{s_2}, \quad D_2 = \frac{\alpha_1}{u_1^2}. \qquad (4.5)$$

Now we note that it is easy to determine the concentration in any pure zone from (4.2). In particular, for zone 2 we have

$$u_1^2 = \alpha_1 \frac{v_2}{\alpha_2}. \qquad (4.6)$$

From (4.5) and (4.6), we find

$$D_2 = \frac{\alpha_2}{v_2}. \qquad (4.7)$$

The solution is conveniently expressed in the form of a table, and the process for searching for the solution is described as filling in the table. The data in Table 4 correspond to the arrangement of zones in Fig. 11.

In the kth column of the table, we write the concentrations $u_1{}^k$, $u_2{}^k$, $u_3{}^k$, $u_4{}^k$ and the jump velocity of the leading edge of the kth zone (in this case, D_k). The presented approaches allowed us to fill in the first three columns. Further filling in of the table is done analogously. From the conditions on the fixed jump D_3, we conclude that zone 4 is pure and contains only component 2; we determine u_2^2 from (4.2) and we determine D_4 from the jump conditions. In this case, it is convenient to choose that condition from conditions (2.33) which corresponds to the component existing on one and only one side of the jump – in this case, $i = 2$ or $i = 3$. We have the equality

$$D_4(u_i^4 - u_i^5) = \alpha_i\left(\frac{u_i^4}{s_4} - \frac{u_i^5}{s_5}\right), \quad i = 1,\ 2,\ 3,\ 4. \tag{4.8}$$

Setting $i = 2$ ($i = 3$) in these equalities and taking into account the fact that $u_2^5 = u_3^4 = 0$ and $u_2^4,\ u_3^5 \neq 0$, we find

$$D_4 = \frac{\alpha_2}{s_4} = \frac{\alpha_2}{u_2^4} = \frac{\alpha_3}{s_5} = \frac{\alpha_3}{u_3^5}. \tag{4.9}$$

Since $u_2^4 = \alpha_2(v_4/\alpha_4)$, from (4.9) we derive

$$D_4 = \frac{\alpha_3}{v_3}, \quad u_3^5 = u_3. \tag{4.10}$$

However, the quantity u_3^5, as in general the concentration in a pure zone, may be determined directly from (4.2). Later, everything will be done analogously.

The arrangement of zones (Fig. 11) and the solution corresponding to it (illustrated in Table 4) are preserved as long as no moving jump crosses a fixed jump. The jump D_2 crosses the fixed jump D_3 at the time $t_{23} = (a_2 - a_1)/D_2 = (a_2 - a_1)v_2/\alpha_2$, and the jump D_4 crosses the fixed jump D_5 at the time $t_{45} = (a_3 - a_2)/D_4 = (a_3 - a_2)v_3/\alpha_3$. After one of these events has occurred, we must replace Fig. 11 with a new figure. For concreteness, let $t_{23} < t_{45}$. Then the zone arrangement illustrated in Fig. 12 occurs.

After passing through D_3, the jump D_2 retains its designation as D_2, but the numbering of the zones changes in connection with the disappearance of the u_2 zone (zone 4 on Fig. 11) and the appearance of the new u_1 zone (zone 3 on Fig. 12). Repeating the preceding approach, we arrive at a solution corresponding to Fig. 12 and represented in Table 5. Here it is noteworthy that the velocities D_2 and D_4 coincide; they are both equal to α_3/v_3. The pure u_1 zone (zone 2) was formed between the fixed jumps and will not change in the following.

The solution illustrated in Table 5 is valid on the time segment t: $t_{23} < t < t_{45}$. When $t > t_{45}$, the jump D_4 reaches the fixed jump D_5, after which

TABLE 5. Concentrations of Substances in the Zones and Jump Velocities Corresponding to Fig. 12

Index of substance (i)	Zone number (k)						
	1	2	3	4	5	6	7
1	v_1	$\alpha_1\dfrac{v_2}{\alpha_2}$	$\alpha_1\dfrac{v_3}{\alpha_3}$	0	0	0	0
2	0	0	0	$\alpha_2\dfrac{v_3}{\alpha_3}$	0	0	0
3	0	0	0	0	v_3	$\dfrac{\alpha_3}{\alpha_4}v_4$	0
4	0	0	0	0	0	0	v_4
Jump velocity for leading edge of kth zone (D_k)	0	0	$\dfrac{\alpha_3}{v_3}$	$\dfrac{\alpha_3}{v_3}$	0	$\dfrac{\alpha_4}{v_4}$	—

TABLE 6. Concentrations of Substances in the Zones and Jump Velocities Corresponding to Fig. 13

Index of substance (i)	Zone number (k)						
	1	2	3	4	5	6	7
1	v_1	$\alpha_1\dfrac{v_2}{\alpha_2}$	$\alpha_1\dfrac{v_3}{\alpha_3}$	0	0	0	0
2	0	0	0	$\alpha_2\dfrac{v_3}{\alpha_3}$	$\alpha_2\dfrac{v_4}{\alpha_4}$	0	0
3	0	0	0	0	0	$\dfrac{\alpha_3}{\alpha_4}v_4$	0
4	0	0	0	0	0	0	v_4
Jump velocity for leading edge of kth zone (D_k)	0	0	$\dfrac{\alpha_3}{v_3}$	0	$\dfrac{\alpha_4}{v_4}$	$\dfrac{\alpha_4}{v_4}$	—

TABLE 7. Concentrations of Substances in the Zones and Jump Velocities Corresponding to Fig. 14

Index of substance (i)	Zone number (k)						
	1	2	3	4	5	6	7
1	v_1	$\alpha_1\dfrac{v_2}{\alpha_2}$	$\alpha_1\dfrac{v_3}{\alpha_3}$	$\alpha_1\dfrac{v_4}{\alpha_4}$	0	0	0
2	0	0	0	0	$\alpha_2\dfrac{v_4}{\alpha_4}$	0	0
3	0	0	0	0	0	$\dfrac{\alpha_3}{\alpha_4}v_4$	0
4	0	0	0	0	0	0	v_4
Jump velocity for leading edge of kth zone (D_k)	0	0	0	$\dfrac{\alpha_4}{v_4}$	$\dfrac{\alpha_4}{v_4}$	$\dfrac{\alpha_4}{v_4}$	—

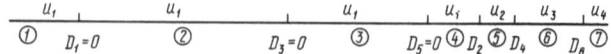

Fig. 14. Final zone arrangement when $t > t_{25}$.

the zones are arranged as shown in Fig. 13 and the solution is illustrated by Table 6. This solution is valid for the times t: $t_{45} < t < t_{25}$; the time at which the jump D_2 reaches the fixed jump D_5 is $t_{25} = (a_2 - a_1)v_2/\alpha_2 + (a_3 - a_2)v_3/\alpha_3$. Finally, for all $t > t_{25}$, the zone arrangement does not change. The final state is described by Fig. 14 and Table 7.

Thus, in the final state the terminal electrolyte forms three fixed zones and one moving zone; each of the rest of the electrolytes are located in their own pure zones, which follow in increasing order of their transport rates. The concentration of each moving u_i zone is determined only by the initial concentration of the leading electrolyte and the transport rate α_i; in this case, $u_i = (\alpha_i/\alpha_4)v_i$. All the moving interfaces (jumps) have the same velocity α_4/v_4, determined only by the initial concentration and the transport velocity of the leader.

Let us now turn to the case of n electrolytes with the initial state determined by condition (4.1), and let us give the rule for filling in the tables, allowing us in all cases to establish a solution.

1. At each instant $t > 0$ (except when there are a finite number of instants of time), a constant number of zones $(2n - 1)$ exist. All these are pure zones; at the initial instant of time, new zones arise as the result of the departure of a moving jump from each initial interface $x = a_i$.

2. The concentration of each pure zone is calculated using formula (4.2) for the invariant R_1. Thus, if the pure u_k zone is located between the points a_{i-1} and a_i, then its concentration is $u_k = (\alpha_k/\alpha_i)v_i$. The concentrations of the leader and the terminal electrolyte (the terminator) in the first zone remain constant.

3. Zones of the same components are located on both sides of a fixed jump.

4. All the zones move to the right. The fixed zones between fixed jumps a_{i-1}, a_i may be formed only by the terminator.

5. All the jumps moving between successive fixed jumps, for example a_{i-1} and a_i (in the region of the initial u_i zone), have the same velocity α_i/v_i.

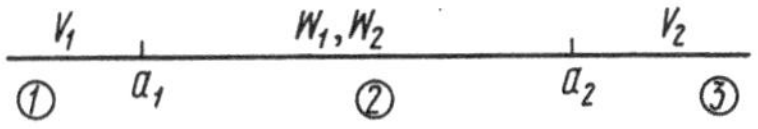

$$\underset{\underset{a_1}{①}}{\underline{\overset{V_1}{\qquad}}} \underset{②}{\underline{\overset{W_1,W_2}{\qquad\qquad}}} \underset{\underset{a_2}{}\ ③}{\underline{\overset{V_2}{\qquad}}}$$

Fig. 15. Initial arrangement of zones for two pure substances, mixed in zone 2.

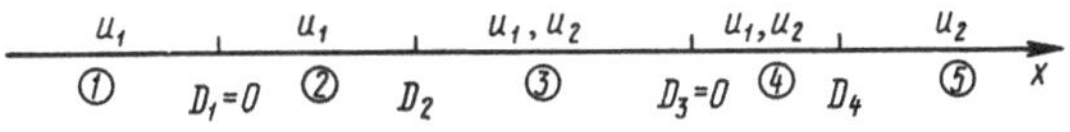

$$\underset{\underset{D_1=0}{①}}{\underline{\overset{u_1}{\qquad}}} \underset{\underset{D_2}{②}}{\underline{\overset{u_1}{\qquad}}} \underset{\underset{D_3=0}{③}}{\underline{\overset{u_1,u_2}{\qquad}}} \underset{\underset{D_4}{④}}{\underline{\overset{u_1,u_2}{\qquad}}} \underset{⑤}{\underline{\overset{u_2}{\qquad}}} \; x$$

Fig. 16. Zones developed from the initial state.

TABLE 8. Concentrations of Substances in the Zones and Jump Velocities Corresponding to Fig. 16

Index of substance (i)	Zone number (k)				
	1	2	3	4	5
1	v_1	$\alpha_1\left(\dfrac{w_1}{\alpha_1} + \dfrac{w_2}{\alpha_2}\right)$	w_1	$\dfrac{v_2/\alpha_2}{w_1/\alpha_1 + w_2/\alpha_2}\, w_1$	0
2	0	0	w_2	$\dfrac{v_2/\alpha_2}{w_1/\alpha_1 + w_2/\alpha_2}\, w_2$	v_2
Jump velocity for leading edge of kth zone (D_k)	0	$\dfrac{\alpha_2}{w_1 + w_2}$	0	$\alpha_1/(K\,(w_1 - w_2))$	—

6. At the instant when a moving jump crosses a fixed jump a_i, the zone adjacent to a_i on the left disappears, and a new zone appears to the right of a_i. It is precisely at these crossing times that an exception to Point 1 is made. Altogether there are $(n-1)(n-2)/2$ crossing times; they are easy to calculate, knowing the jump velocities in the regions between the fixed jumps. For example, the last instant of time $t_{2,n-1}$ for the crossing of the jump departing from the initial jump a_1 over the last fixed jump a_{n-1} is determined by the equality $t_{2,n-1} = (a_2 - a_1)v_2/\alpha_2 + (a_3 - a_2)v_3/\alpha_3 + \ldots + (a_{n-1} - a_{n-2})v_{n-1}/\alpha_{n-1}$.

7. After all the moving jumps have passed through the interval (a_{i-1}, a_i), in this interval a fixed terminator zone is formed (of course, the movement of ions through the zone continues).

8. For $t > t_{2,n-1}$, when all the moving jumps generated by the initial jumps have already passed through all the fixed jumps, the final state is formed in which the terminator forms $(n-1)$ fixed zones and one moving zone, and the rest of the electrolytes are arranged in increasing order of their transport rates; the ith electrolyte ($i = 2, \ldots, n-1$) forms one moving

pure u_i zone with concentration $\alpha_i(v_n/\alpha_n)$ $(i = 1, 2, ..., n)$, and the leading edges move with the same velocity v_n/α_n.

We note that the solution is given in u_i variables; transition to the quantities ξ_i is accomplished using formula (1.10).

5. Case of Two Partially Mixed Electrolytes

Let an infinite column be filled with two electrolytes and at the initial instant of time let there exist two infinite pure zones 1 and 3 (Fig. 15) with concentrations, respectively, $u_1 = v_1$ and $u_2 = v_2$, separated by a mixed zone with concentrations $u_1 = w_1$, $u_2 = w_2$. Thus the initial conditions for system (2.1) for $n = 2$ have the form

$$u_1 = \begin{cases} v_1, & x < a_1, \\ w_1, & a_1 < x < a_2, \\ 0, & x > a_2, \end{cases} \qquad u_2 = \begin{cases} 0, & x < a_1, \\ w_2, & a_1 < x < a_2, \\ v_2, & x > a_2. \end{cases} \qquad (5.1)$$

We then need to establish the generalized piecewise-constant solution of the Cauchy problem. As in the preceding section, we assume that $\alpha_2 > \alpha_1$, i.e., the fast electrolyte is located "downstream," and we represent the solution by means of a series of figures with the zone arrangements and tables of concentrations in the zones as well as the jump velocities. As before, going to the next figure will correspond to crossing of a fixed jump by one of the moving jumps.

There are two fixed jumps in this problem: $x = a_1$ and $x = a_2$. The condition that the Riemann invariant $R_1 = (u_1/\alpha_1) + (u_2/\alpha_2)$ is constant gives the equality

$$R_1 = \begin{cases} \dfrac{v_1}{\alpha_1}, & x < a_1, \\ \dfrac{w_1}{\alpha_1} + \dfrac{w_2}{\alpha_2}, & a_1 < x < a_2, \\ \dfrac{v_2}{\alpha_2}, & x > a_2. \end{cases} \qquad (5.2)$$

The solution describing the initial course of the process is illustrated in Fig. 16, where the zone arrangement is presented, and in Table 8, where the kth column contains the concentrations u_i^k in the kth zone and the velocity of the jump forming its leading edge. The moving jumps D_2 and D_4

start from the fixed jumps D_1 and D_3. The fact that zone 2 is a pure u_1 zone while zone 4 is a mixed zone is easily derived from the conditions on the fixed jump. In fact, the latter require that the quantities u_1/s be continuous on the fixed jumps.

Let us describe how to fill in Table 8. First of all, we note the velocities of the fixed jumps ($D = 0$) in columns 1 and 3. Accordingly, the concentrations are transferred to these columns from the initial condition (5.1). The same also applies to the last column. In the leader zone, the concentrations do not change. D is missing in the last column; there is no leading edge in the leader zone.

Let us go on to filling in column 2 of Table 8. We determine the concentration u_1 from the condition that the Riemann invariant R_1 is constant [see (5.2)]: since $u_2^2 = 0$, we have $u_1 = \alpha_1 R_1 = \alpha_1[(w_1/\alpha_1) + (w_2/\alpha_2)]$. From the conditions on the moving jump D_2,

$$D_2 (u_i^2 - u_i^3) = \alpha_i \left(\frac{u_i^2}{s_2} - \frac{u_i^3}{s_3} \right) \quad (i = 1,\ 2),\tag{5.3}$$

we find, setting $i = 2$ and taking into account the fact that $u_2^2 = 0$, $u_2^3 \neq 0$,

$$D_2 = \frac{\alpha_2}{s_3} = \frac{\alpha_2}{w_1 + w_2}.\tag{5.4}$$

Of course, the same result is also obtained from (5.3) when $i = 1$. As has already been indicated, the answer is obtained faster if we use the jump condition which corresponds to the component existing only on one side of the jump.

We still need to fill in column 4. From the condition for the component on the moving jump D_4, we obtain

$$D_4 = \frac{\alpha_1}{s_4}.\tag{5.5}$$

We obtain the system of equations for determining u_1^4, u_2^4 using the constancy of the Riemann invariant R_1: equality (5.2) and the continuity of u_1/s on the fixed jump D_3. Thus, we have

$$\frac{u_1^4}{\alpha_1} + \frac{u_2^4}{\alpha_2} = \frac{v_2}{\alpha_2}, \qquad \frac{u_1^4}{s_4} = \frac{u_1^3}{s_3} = \frac{w_1}{w_1 + w_2}, \qquad s_4 = u_1^4 + u_2^4.\tag{5.6}$$

Solving system (5.6), we find

$$u_1^4 = \frac{v_2/\alpha_2}{w_1/\alpha_1 + w_2/\alpha_2}\, w_1, \qquad u_2^4 = \frac{v_2/\alpha_2}{w_1/\alpha_1 + w_2/\alpha_2}\, w_2.\tag{5.7}$$

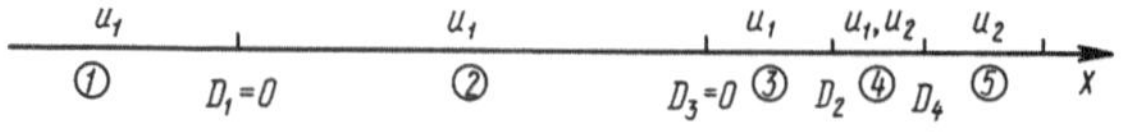

Fig. 17. Arrangement of zones in the time interval $t_{23} < t < t_{24}$.

TABLE 9. Concentrations of Substances in the Zones and Jump Velocities Corresponding to Fig. 17

Index of substance (i)	Zone number (k)				
	1	2	3	4	5
1	v_i	$\alpha_1\left(\dfrac{w_1}{\alpha_1} + \dfrac{\alpha_2}{w_2}\right)$	$\alpha_1\dfrac{v_2}{\alpha_2}$	$\dfrac{v_2/\alpha_2}{w_1/\alpha_1 + w_2/\alpha_2}w_1$	0
2	0	0	0	$\dfrac{v_2/\alpha_2}{w_1/\alpha_1 + w_2/\alpha_2}w_2$	v_2
Jump velocity for leading edge of kth zone (D_k)	0	0	$\dfrac{\alpha_2}{K(w_1 + w_2)}$	$\alpha_1/K(w_1 + w_2)$	—

We note that the fraction in (5.7) is the ratio of the invariant R_1 to the right and to the left of the fixed jump D_3.

Finally, we determine the velocity D_4 from the condition on the moving jump D_4 for the component u_1:

$$D_4 = \frac{\alpha_i}{s_4} = \frac{\alpha_i}{K(w_1 + w_2)}, \quad K = \frac{R_1^4}{R_1^3} = \frac{v_2/\alpha_2}{w_1/\alpha_1 + w_2/\alpha_2}. \qquad (5.8)$$

The solution found is valid as long as the jump D_2 does not cross the fixed jump D_3, which occurs at the time

$$t_{23} = \frac{a_2 - a_1}{D_2} = \frac{(a_2 - a_1)(w_1 + w_2)}{\alpha_1}. \qquad (5.9)$$

After this time, a new distribution of zones occurs. Zone 3 in Fig. 16 "disappears" and a new zone appears: zone 3 in Fig. 17. The solution is now illustrated by Table 9, which is filled out in the following order. Columns 1, 2, 4, and 5 are transferred from Table 8; the changes do not affect the "old" zones. Zone 3, as a neighbor to the pure u_1 zone 2 relative to the fixed jump D_3, is itself a pure u_1 zone, and u_1 is determined from the invariant R_1. We determine the velocity D_2 from the jump condition for the component u_2, and as a result we find

$$D_2 = \frac{\alpha_2}{s_4} = \frac{\alpha_2}{K(w_1 + w_2)}. \qquad (5.10)$$

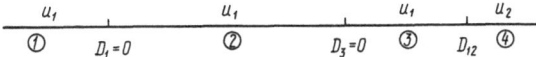

Fig. 18. Final arrangement of zones when $t > t_{24} = t_*$.

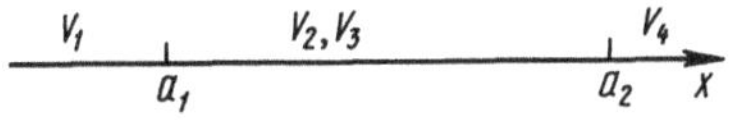

Fig. 19. Initial arrangement of zones.

TABLE 10. Concentrations of Substances in the Zones
and Jump Velocities Corresponding to Fig. 18

Index of substance (i)	Zone number (k)			
	1	2	3	4
1	v_1	$\alpha_1\left(\dfrac{w_1}{\alpha_1}+\dfrac{w_2}{\alpha_2}\right)$	$\alpha_1\dfrac{v_2}{\alpha_2}$	0
2	0	0	0	v_2
Jump velocity for leading edge of kth zone (D_k)	0	0	$\dfrac{\alpha_2}{v_2}$	—

The established solution is in force as long as the jump D_2 does not reach the jump D_4; since $D_2 > D_4$ (by virtue of the assumption that $\alpha_2 > \alpha_1$), this in fact occurs at the instant of time

$$t_* = t_{23} + \frac{D_4}{D_2 - D_4}\, t_{23} = \frac{\alpha_1}{\alpha_2 - \alpha_1}\, t_{23} = \frac{(a_2 - a_1)\,(w_1 + w_2)}{\alpha_2 - \alpha_1}\,. \quad (5.11)$$

The quantity t_* gives the time for complete separation of the electrolytes. As is evident from (5.11), it is proportional to the total mass of electrolytes in the initial mixed zone (the numerator) and is inversely proportional to the difference in transport rates.

When $t > t_*$, we have the final distribution (Fig. 18): the electrolytes are completely separated, and the leader is found in a single moving zone, while the terminator forms two fixed zones (the ions move!) and one moving zone. The interface between the moving u_1 and u_2 zones travels at the velocity D_{12}, which is easily determined from the jump conditions:

$$D_{12} = \frac{\alpha_1}{u_1^3} = \frac{\alpha_2}{u_1^4} = \frac{\alpha_2}{v_2}\,. \quad (5.12)$$

The velocity of the interface, as would be expected, depends only on the transport rate of the leader and its initial concentration in the pure zone.

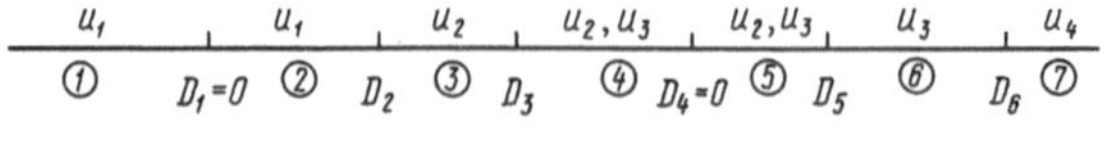

Fig. 20. Zones developing from the initial state.

TABLE 11. Concentrations of Substances in the Zones and Jump Velocities Corresponding to Fig. 20

Index of substance (i)	Zone number (k)						
	1	2	3	4	5	6	7
1	v_1	$\alpha_1\left(\dfrac{v_1}{\alpha_1}+\dfrac{v_2}{\alpha_2}\right)$	0	0	0	0	0
2	0	0	$\alpha_2\left(\dfrac{v_2}{\alpha_2}+\dfrac{v_3}{\alpha_3}\right)$	v_2	Kv_2	0	0
3	0	0	0	v_3	Kv_3	$\dfrac{\alpha_3}{\alpha_4}v_4$	0
4	0	0	0	0	0	0	v_4
Jump velocity for leading edge of kth zone (D_k)	0	$\dfrac{1}{\dfrac{v_2}{\alpha_2}+\dfrac{v_3}{\alpha_3}}$	$\dfrac{\alpha_3}{v_2+v_3}$	0	$\dfrac{\alpha_2}{K\,(v_2+v_3)}$	$\dfrac{\alpha_4}{v_4}$	—

The final state (the solution after complete separation, i.e., when $t > t_*$) is described by Table 10. We note that Table 10 may be constructed independently of the calculations in this section. All the zones are pure, and their concentrations are determined from the invariant R_1.

6. Separation of Two-Component Mixtures by the Isotachophoresis Method

At the initial instant of time $t = 0$, let the column contain terminator and leader zones with constant concentrations (respectively v_1 and v_4), and let the mixture to be separated, composed of two electrolytes with concentrations v_2 and v_3, be placed between these two zones. The development of the process is described by the Cauchy problem for a system (2.1) with the initial condition illustrated in Fig. 19. We will assume that the transport rates satisfy the condition $\alpha_1 < \alpha_2 < \alpha_3 < \alpha_4$, which ensures the existence of a piecewise-constant generalized solution to the problem (the jump stability condition is satisfied).

The arrangement of the zones at the beginning of the process is illustrated in Fig. 20, and the solution is written in Table 11. Let us describe the order for filling in this table. We transfer to the columns 1, 4, and 7 the

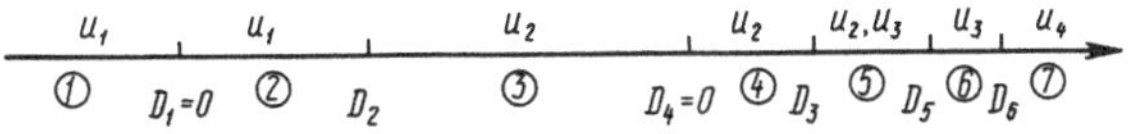

Fig. 21. Arrangement of zones in the time interval $t_{34} < t < t_{24}$.

TABLE 12. Concentrations of Substances in the Zones and Jump Velocities Corresponding to Fig. 21

Index of substance (i)	Zone number (k)						
	1	2	3	4	5	6	7
1	v_1	$\alpha_1\left(\dfrac{v_2}{\alpha_2}+\dfrac{v_3}{\alpha_3}\right)$	0	0	0	0	0
2	0	0	$\alpha_2\left(\dfrac{v_2}{\alpha_2}+\dfrac{v_3}{\alpha_3}\right)$	$\alpha_2\dfrac{v_4}{\alpha_4}$	Kv_2	0	0
3	0	0	0	0	Kv_3	0	0
4	0	0	0	0	0	0	0
Jump velocity for leading edge of kth zone (D_k)	0	1 $\dfrac{v_2}{\alpha_2}+\dfrac{v_3}{\alpha_3}$	0	$\dfrac{\alpha_3}{v_2+v_3}$	$\dfrac{\alpha_2}{K\,(v_2+v_3)}$	$\dfrac{\alpha_4}{v_4}$	—

initial concentrations and the zero jump velocities D_1 and D_2. We determine the concentrations in zone 2, using the continuity of the quantity u_i/s on the fixed jump D_1 and the constancy of R_1, since $u_1^3 = 0$ and the slow electrolyte of zone 2 cannot mix with the electrolytes of zone 3 (this follows directly from consideration of the characteristics; see Section 3). We determine the velocity from the condition on this jump for the u_2 component

$$D_2 = \frac{\alpha_2}{s_3} = \frac{1}{\dfrac{v_2}{\alpha_2}+\dfrac{v_3}{\alpha_3}}. \qquad (6.1)$$

We determine the velocity D_3 from the conditions on this jump for the u_3 component

$$D_3 = \frac{\alpha_3}{s_4} = \frac{\alpha_3}{v_2+v_3}. \qquad (6.2)$$

Let us go on to zone 6. Since according to the properties of the characteristics of Eq. (2.1) "the leader only goes forward," we have $u_4^6 = 0$. Then, from the conditions on the jump D_6 we conclude that $u_1^6 = u_2^6 = 0$, and D_6 is determined by the formula

$$D_6 = \frac{\alpha_3}{s_6} = \frac{\alpha_4}{v_4}. \qquad (6.3)$$

Here we have already used the value $u_3^6 = (\alpha_3/\alpha_4)v_4$, which is determined from the condition that R_1 is constant, since zone 6 is a pure zone.

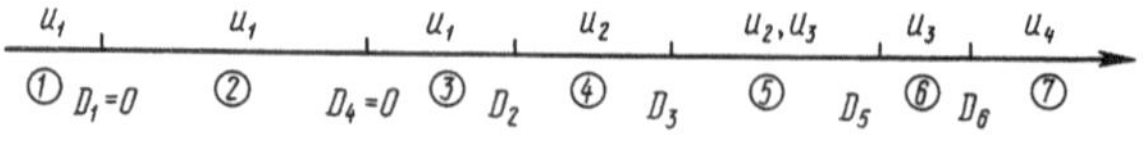

Fig. 22. Arrangement of zones in the time interval $t_{24} < t < t_{35}$.

TABLE 13. Concentrations of Substances in the Zones and Jump Velocities Corresponding to Fig. 22

Index of substance (i)	Zone number (k)						
	1	2	3	4	5	6	7
1	v_1	$\alpha_1\left(\dfrac{v_2}{\alpha_2}+\dfrac{v_3}{\alpha_3}\right)$	$\alpha_1\dfrac{v_4}{\alpha_4}$	0	0	0	0
2	0	0	0	$\dfrac{\alpha_2}{\alpha_4}v_4$	Kv_2	0	0
3	0	0	0	0	Kv_3	$\dfrac{\alpha_3}{\alpha_4}v_4$	0
4	0	0	0	0	0	0	v_4
Jump velocity for leading edge of kth zone (D_k)	0	0	$\dfrac{\alpha_4}{v_4}$	$\dfrac{\alpha_3}{K\,(v_2+v_3)}$	$\dfrac{\alpha_2}{K\,(v_2+v_3)}$	$\dfrac{\alpha_4}{v_4}$	—

We still need to fill in column 5. By virtue of the finiteness of the velocity of disturbances for the hyperbolic system, here $u_1{}^5 = 0$; it has already been made clear above that we also have $u_4{}^5 = 0$. From the condition for u_2 on the jump D_5, also using the constancy of R_1, we obtain

$$D_5 = \frac{\alpha_2}{s_5}, \qquad (6.4)$$

$$\frac{u_2^5}{\alpha_2} + \frac{u_3^5}{\alpha_3} = \frac{v_4}{\alpha_4}. \qquad (6.5)$$

The continuity condition for u_2/s on the fixed jump D_4 gives the equality

$$\frac{u_2^4}{s_4} = \frac{u_2^5}{s_5}. \qquad (6.6)$$

Solving system (6.4)-(6.6), we find

$$u_2^5 = Kv_2, \; u_3^5 = Kv_3, \; K = \frac{v_4/\alpha_4}{v_2/\alpha_2 + v_3/\alpha_3}, \qquad (6.7)$$

$$D_5 = \frac{\alpha_2}{K\,(v_2+v_3)}. \qquad (6.8)$$

We note that $D_2 < D_4$ and $D_5 < D_6$, since the jump D_2 cannot reach the jump D_3, and the jump D_5 cannot reach the jump D_6. Therefore, the

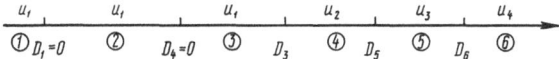

Fig. 23. Final state when $t > t_{35} = t_*$.

TABLE 14. Concentrations of Substances in the Zones and Jump Velocities Corresponding to Fig. 23

Index of substance (i)	Zone number (k)					
	1	2	3	4	5	6
1	v_1	$\alpha_1\left(\dfrac{v_2}{\alpha_2}+\dfrac{v_3}{\alpha_3}\right)$	$\alpha_1\dfrac{v_4}{\alpha_4}$	0	0	0
2	0	0	0	$\alpha_2\dfrac{v_4}{\alpha_4}$	0	0
3	0	0	0	0	$\dfrac{\alpha_3}{\alpha_4}v_4$	0
4	0	0	0	0	0	v_4
Jump velocity for leading edge of kth zone (D_k)	0	0	$\dfrac{\alpha_4}{v_4}$	$\dfrac{\alpha_4}{v_4}$	$\dfrac{\alpha_4}{v_4}$	—

first zone rearrangement (disappearance of zone 4 on Fig. 20 and appearance of zone 4 on Fig. 21) occurs when the jump D_3 reaches the fixed jump D_4. The time

$$t_{34} = \frac{a_2 - a_1}{D_3} = \frac{(a_2 - a_1)(v_2 + v_3)}{\alpha_3} \tag{6.9}$$

corresponds to this event. It is curious that the quantity t_{34} does not depend on α, provided that α_2 is less than α_3.

The new arrangement of zones is illustrated in Fig. 21, and the solution is presented in Table 12. All the columns of Table 12 except columns 3 and 4 are transferred from Table 11 without change. In the last row of the third column we write 0; now zone 3 is bounded in front by the fixed jump D_4. The concentrations in column 3 remain the same as in Table 11. The concentrations in zone 4 are determined using the conditions on the jumps D_4 and D_3 by repeated application of the described method; the new jump velocity D_3 is more rapidly determined from the condition for u_3 on this jump. We obtain

$$D_3 = \frac{\alpha_3}{s_5} = \frac{\alpha_3}{K(v_2 + v_3)}. \tag{6.10}$$

The following rearrangement of the zones corresponds to the times t_{24} (when D_2 reached D_4) and t_{35} (when D_3 reached D_5). In this case we have

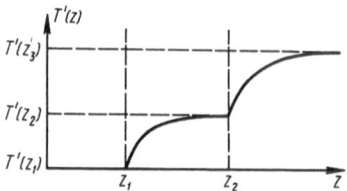

Fig. 24. Form of the function $T'(z)$, proportional to the temperature difference between the zones.

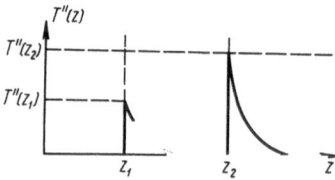

Fig. 25. Form of the function $T''(z)$, corresponding to the derivative signal registered by the recorder.

$$t_{24} = \frac{a_2 - a_1}{D_2} = (a_2 - a_1)\left(\frac{v_2}{\alpha_2} + \frac{v_3}{\alpha_3}\right),$$

$$t_{35} = t_{34} + \frac{D_5}{D_3 - D_5} t_{34} = \frac{D_3}{D_3 - D_5} t_{34} = \frac{1}{\alpha_3 - \alpha_2}(a_2 - a_1)(v_2 + v_3) \tag{6.11}$$

Which arrangement occurs first depends on whether $t_{24} < t_{35}$ or $t_{24} > t_{35}$. As an example, let us consider the first case. There the new arrangement of zones has the form shown on Fig. 22, and the solution is represented by Table 13. Table 13 differs from Table 12 in column 3, where now we write the new value of D_3,

$$D_3 = \frac{\alpha_1}{s_3} = \frac{\alpha_4}{v_4}, \tag{6.12}$$

and in column 2, where now $D = D_4 = 0$.

The last rearrangement and complete separation begins at the time t_*; in this case, $t_* = t_{35}$ [see (6.11)]. The final state is described by Fig. 23 and Table 14. In the case $t_{24} > t_{35}$, the separation has already begun at the time $t = t_{35}$. Figure 23 and Table 14 correspond to this case, i.e., for $t > t_{35}$, the final state begins; this state is certainly the same in both cases and, in general, is calculated independently of the intermediate computations presented.

7. Temperature Distribution for Completely Separated Zones

The method of registering the boundaries between sample zones is of great importance in isotachophoresis practice. The thermometric method [46, 95, 76] is especially used for this purpose. The principles of this method involve the following. The concentration of the electrolytes in different zones is constant but not the same; consequently, the electrical conductivity of the mixture is piecewise-constant along the electrophoretic chamber. In order to maintain a constant electric current density through the mixture, the electric field intensity also should be piecewise-constant. As a result, the heat evolved by the Joule effect in different zones is not the same and, consequently, the temperatures of the zones are different.

We may define the interface by measuring the temperatures of the different zones. In practice, a thermocouple registering the temperature jumps is attached to the electrophoretic column (the chamber); as the zone boundaries pass across the thermocouple, a temperature jump occurs (for convenience in registration, the signal coming from the thermocouple may be differentiated).

In this section, we give a calculation for the temperature field in the electrophoretic column when complete zone separation occurs. We may calculate the temperature field in the general case in a similar manner.

The thermal conductivity equation has the form

$$T_t - \chi T_{xx} = \delta_0 (Ej - Q), \qquad (7.1)$$

where Q is the capacity of the cooling device.

Let us consider the case where complete zone separation occurs: all the zones travel at the same velocity c. Going to a coordinate system which moves along with the zones,

$$z = x - ct, \qquad (7.2)$$

we obtain

$$-cT' - \chi T'' = \delta_0 (Ej - Q), \quad (\ )' \equiv \frac{\partial (\ )}{\partial z}. \qquad (7.3)$$

If all the zones at the initial instant of time are "pure" (see Sections 4-6), the heat source has the form

$$\delta_0 (Ej - Q) \equiv \delta_0 \sum_{k=1}^{n} a_k h (z - z_k), \quad h (z - z_k) = \begin{cases} 1, & z \geq z_k, \\ 0, & z < z_k, \end{cases} \qquad (7.4)$$

where z_k is the boundary between the zones in the moving coordinate system, $(a_{k+1} - a_k)$ are the jumps of the heat source ($k = 0, 1, ..., n - 1$; $a_0 \equiv 0$), easily determined from the results of Sections 3-6; $h(z - z_k)$ is the Heaviside function.

For convenience, let us introduce the symbol

$$c_0 = \frac{c}{\chi} . \tag{7.5}$$

Integrating (7.3) from $-\infty$ to z, and taking into account (7.4) and (7.5), we obtain

$$e^{-c_0 z} (e^{c_0 z}T)' = -\frac{\delta_0}{\chi} \sum_{k=1}^{n} \int_{-\infty}^{z} h (s - z_k) \, ds = -\frac{\delta_0}{\chi} \sum_{k=1}^{n} a_k h_k (z - z_k)$$
$$\times (z - z_k). \tag{7.6}$$

Multiplying the expression obtained by $e^{c_0 z}$ and integrating from $-\infty$ to z, we derive an expression for the temperature:

$$T (z) = -\frac{\delta_0}{\chi} \sum_{k=1}^{n} a_k h (z - z_k) \frac{1}{c_0} \left\{ (z - z_k) - \frac{1}{c_0} [1 - e^{-c_0(z - z_k)}] \right\} . \tag{7.7}$$

From this we have

$$T' (z) = -\frac{\delta_0}{\chi} \sum_{k=1}^{n} a_k x (z - z_k) \frac{1}{c_0} \{ 1 - e^{-c_0(z - z_k)} \}, \tag{7.8}$$

$$T'' (z) = -\frac{\delta_0}{\chi} \sum_{k=1}^{n} a_k h (z - z_k) e^{-c_0(z - z_k)}. \tag{7.9}$$

The form of the functions $T'(z)$, $T''(z)$ is shown in Figs. 24 and 25.

The quantity $T'(z)$ is registered by the thermocouple [the temperature difference between zones for a fixed thermocouple length Δl is equal to $\Delta T = T'(z) \cdot \Delta l$.] The function $T''(z)$ is the derivative of the signal registered by the recorder. For reliable registration of the zone interfaces, it is necessary that the signals $T''(z)$ coming from different boundaries do not overlap one another.

Let us introduce the effective width of the signal δ, the distance at which the signal intensity drops by a factor of e:

$$\delta = \frac{1}{c_0} \, . \tag{7.10}$$

Then the necessary condition for reliability of registration of the boundaries of the moving zones z_k and z_{k+1} takes on the form

$$\delta \ll |z_{k+1} - z_k| \, . \tag{7.11}$$

Taking into account (7.5) and (7.10), we have

$$\frac{\chi}{c} \ll |z_{k+1} - z_k| . \tag{7.12}$$

Thus, the thermometric registration method is applicable for weakly thermally conducting solutions and for rapidly moving zones. Let us recall that the zone velocity may be increased by increasing the electric current density. However, this is undesirable since, in this case, the heat evolution due to the Joule effect increases, and consequently the temperature of the mixture also increases.

Chapter V

MODEL OF ZONE ELECTROPHORESIS

At the present time, zone electrophoresis refers to a number of the most widely used methods of separation, purification, and analysis of biopolymers. As has been indicated in Chapter III, various artificial density gradients, supporting media, etc., have been used with the goal of stabilizing the process relative to gravitational convection. Free liquid-zone electrophoresis in practice has not been applied in pure form [33, 57, 63]. Nevertheless, the essential features of the process may be understood and explained in the free liquid electrophoresis model.

The idea of zone electrophoresis is simple and obvious: at the initial instant of time, the mixture may be represented as a collection of spatially superimposed samples, each of which begins to move under the influence of an electric field at its own constant velocity, determined by the mobility or the transport rate of the charged particles (ions or zwitterions).

The great advantage of this method is the speed of separation. However, we must bear in mind that, in contrast to isoelectric focusing, for example, here there are no factors which promote the concentration of the sample. Without taking into account diffusion, the width of each zone would remain constant; diffusion only increases the zone width. Therefore, it is expedient to continue the process only up to clear zone separation.

The above is valid if the concentration of the samples is small enough that we may neglect the reaction of the samples with each other and with the buffer. The nonreacting zone model is presented in Section 1 of this chapter, where we present an asymptotic solution demonstrating the broadening of the moving zone under the influence of diffusion.

In Section 2, we consider a model describing the reaction between the sample zone and the buffer. It is shown that, even in the absence of diffusion, a mechanism exists which ultimately leads to unlimited growth in the zone width. This effect, induced by the appearance of a rarefaction wave, was called electrolytic mixing in Chapter IV. We should emphasize the similarity between this model and the isotachophoresis model. We also note the fact (not previously noted in the literature) that constriction of the zone occurs in the initial section of the process; this once again points to the need for concluding the process at an opportune moment.

In this chapter we also consider thermal effects accompanying zone electrophoresis. The theory given in Section 3 describes the distortion of the zone profile in the electrophoretic column due to the radial steady-state temperature distribution, a phenomenon which is well known to experimenters.

1. Electrophoresis in an Infinite Column

In this section, we consider the simplest model for zone electrophoresis. Let the mixture of chemical subsystems be located in an electrophoretic column whose length is significantly greater than its inner diameter. We will look for a cylindrically symmetric solution to system (III.5.16)-(III.5.20). In this case, the problem is reduced to the one-dimensional problem:

$$\frac{\partial \xi_1}{\partial t} + \frac{\partial}{\partial x} \left\{ - \mu D_4 \frac{\partial \xi_1}{\partial x} + \beta e^{-\psi} \operatorname{sh}(\psi - \psi^i) \xi_1 \right\} = 0, \qquad (1.1)$$

$$\frac{\partial \xi_2}{\partial t} - \mu D_2 \frac{\partial^2 \xi_2}{\partial x^2} = 0, \qquad (1.2)$$

$$e^{\psi} = \sqrt{\xi_2}, \quad E = \frac{1}{\sqrt{\xi_2}(1 + D_3)} \cdot j, \quad \frac{\partial j}{\partial x} = 0, \quad \beta = \frac{jD}{m(1 + D_3)}. \qquad (1.3)$$

Considering the electrophoretic column to be infinite, let us specify the following boundary and initial conditions:

$$\left\{ - \mu D_4 \frac{\partial \xi_1}{\partial x} + \beta e^{-\psi} \operatorname{sh}(\psi - \psi^i) \xi_1 \right\}\Big|_{x=0,\infty} = 0, \qquad (1.4)$$

$$\xi_2 |_{x=0,\infty} = \tilde{\xi}_2 = \text{const}, \qquad (1.5)$$

$$j |_{x=0,\infty} = J(t), \qquad (1.6)$$

$$\xi_1 |_{t=0} = \tilde{\xi}_1(x), \qquad (1.7)$$

$$\xi_2 \mid_{t=0} = \tilde{\xi}_2 = \text{const } (e^{\psi} \mid_{t=0} = \text{const}). \tag{1.8}$$

These conditions correspond to the facts that the pH of the solution is maintained at the specified constant level, the total amount of sample in the electrophoretic column is constant, and the external current is specified. It is easy to see that in this case system (1.2), (1.3) with conditions (1.5), (1.6), (1.8) has the solution

$$\xi_2(x, t) = \tilde{\xi}_2, \quad \psi(x, t) \equiv \frac{1}{2} \ln \tilde{\xi}_2 = \text{const},$$

$$E(x, t) \equiv E(t) = \frac{1}{\sqrt{\tilde{\xi}_2} (1 + D_3)} J(t), \quad j(x, t) = J(t). \tag{1.9}$$

This solution corresponds to the case where the buffer concentration ξ_2 and the acidity ψ within the electrophoretic column do not change over the course of time and are equal to their values at the ends of the columns ($x = 0$, $x = \infty$), and the electric current density and electric field intensity are specified functions of time.

Now we write Eq. (1.1) with conditions (1.4), (1.7) in the form (for convenience, we omit the subscript 1: $\xi_1 \equiv \xi$)

$$\frac{\partial \xi}{\partial t} + \frac{\partial}{\partial x} \left\{ -\mu_0 \frac{\partial \xi}{\partial x} + \beta_0 \xi \right\} = 0, \tag{1.10}$$

$$\left(-\mu_0 \frac{\partial \xi}{\partial x} + \beta_0 \xi \right) \Big|_{x=0,\infty} = 0, \tag{1.11}$$

$$\xi \mid_{t=0} = \tilde{\xi}(x), \tag{1.12}$$

where

$$\mu_0 \equiv \mu D_4, \quad \beta_0 = \beta_0(t) \equiv \frac{D \, \text{sh} \left(\frac{1}{2} \ln \tilde{\xi}_2 - \psi^i \right)}{m (1 + D_3) \sqrt{\tilde{\xi}_2}} J(t). \tag{1.13}$$

The simplest case, which is most frequently realized in practice, is the case where $J(t) \equiv \text{const}$. In this case,

$$\beta_0 = \text{const.} \tag{1.14}$$

Let us restrict ourselves to consideration of this case (for the other cases, see the end of Section 1, Chapter IX). Equation (1.10) is the diffusion equation with transport (β_0 is the transport rate). We will look for its solution in the form

$$\xi(x,\ t) \equiv \xi(x - \beta_0 t_L,\ t_L),\quad t = t_L,\quad z = x - \beta_0 t_L, \tag{1.15}$$

where t_L is the time measured in the moving coordinate system. Without reducing generality, condition (1.11) may be replaced by the following:

$$\xi|_{z=\pm\infty} = 0 \ \left(\frac{\partial \xi}{\partial z}\bigg|_{z=\pm\infty} = 0 \right). \tag{1.16}$$

Equation (1.10) in z, t_L variables has the form

$$\frac{\partial \xi}{\partial t_L} - \mu_0 \frac{\partial^2 \xi}{\partial z^2} = 0,\quad -\infty < z < \infty. \tag{1.17}$$

The solution to (1.17), taking into account the initial condition (1.12), is written in the form

$$\xi(x,\ t) \equiv \xi(x - \beta_0 t_L,\ t_L) = \frac{1}{2\sqrt{\pi}} \int\limits_{-\infty}^{\infty} \frac{1}{\sqrt{\mu_0 t}}\, e^{-\frac{(x - \beta_0 t - s)^2}{4\mu_0 t}} \cdot \tilde{\xi}(s)\, ds. \tag{1.18}$$

Thus, the initial concentration profile $\tilde{\xi}(x)$ moves at the velocity β_0, broadened as a result of diffusion. If the following condition is satisfied:

$$\mu_0 t \ll 1, \tag{1.19}$$

then expression (1.18), using the Laplace method (see [53]), is rewritten in the form

$$\xi(x,\ t) = \sum_{k=0}^{\infty} \frac{1}{(2k)!}\, \frac{\Gamma\left(k + \frac{1}{2}\right)(4\mu_0 t)^{k+\frac{1}{2}}}{2\sqrt{\mu_0 \pi t}} \left\{ \frac{d^{2k}}{dz^{2k}}\, \tilde{\xi}(z) \right\}\bigg|_{z=x-\beta_0 t}. \tag{1.20}$$

Restricting ourselves to the first two terms, we obtain

$$\xi(x,\ t) = \tilde{\xi}(x - \beta_0 t) + \mu_0 t \tilde{\xi}''(x - \beta_0 t) + O(\mu_0^2 t^2),\quad \mu_0 t \to 0. \tag{1.21}$$

Let us recall [see (II.3.2), (II.3.3)] that the condition (1.19) in dimensional variables has the form

$$t_d \ll \frac{L^2}{D_{4,d}}, \tag{1.22}$$

where t_d is the dimensional time; $D_{4,d}$ is the dimensional diffusion coefficient for the sample; L is the characteristic length. Thus, for times significantly less than the characteristic diffusion length $L^2/D_{4,d}$, formula (1.21) is a good approximation to the solution.

2. Reaction of the Zones with the Buffer

The simplest model of zone electrophoresis, considered in the preceding section, does not take into account the effect exerted by the moving zone on the buffer solution. This essentially is part of the basis of the model: the sample concentration is chosen to be negligibly small compared with the buffer concentration. Such a model can only describe the change in the concentration profile of the moving zone as a result of its broadening by diffusion.

In this section, we consider a zone electrophoresis model for the case where the concentrations of the components of the buffer and samples are comparable with one another and diffusion is negligibly small. In this case, it is reasonable to consider that the change in the concentration profile of the samples will mainly be determined by the reactions of the samples with the buffer. Such a model is reminiscent of the isotachophoresis model: as we will see in the following development, the system of equations describing the process coincides with (IV.2.1). Mathematically, the difference involves specification of the initial conditions; the samples are surrounded by the buffer solution.

Let us assume that there is no diffusion in the mixture:

$$\mu = 0. \tag{2.1}$$

Let the mixture be enclosed in an infinite electrophoretic column. Then, in the one-dimensional case, system (IV.2.1) takes on the form

$$\frac{\partial \xi_k}{\partial t} + \frac{\partial}{\partial x} \{\gamma_k \xi_k E\} = 0, \quad k = 1, \ldots, n, \tag{2.2}$$

$$a_1 + \sum_{i=1}^{n} e_i \xi_i = 0, \tag{2.3}$$

$$j = E\left(a_1 + \sum_{i=1}^{n} \sigma_i \xi_i\right). \tag{2.4}$$

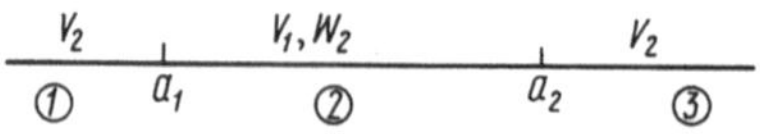

$$\frac{\quad V_2 \quad | \quad V_1, W_2 \quad | \quad V_2 \quad}{① \quad a_1 \quad ② \qquad a_2 \quad ③}$$

Fig. 26. Initial arrangement of zones.

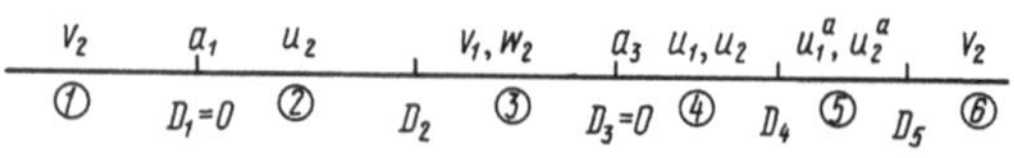

$$\frac{\quad V_2 \quad a_1 \quad u_2 \quad | \quad V_1, W_2 \quad a_3 \ u_1, u_2 \ u_1^a, u_2^a \quad V_2 \quad}{① \quad D_1{=}0 \ ② \quad D_2 \quad ③ \quad D_3{=}0 \ ④ \quad D_4 \ ⑤ \quad D_5 \ ⑥}$$

Fig. 27. Zones developed from the initial state.

TABLE 15. Concentrations of Substances in the Zones and Jump Velocities Corresponding to Fig. 27

Index of substance (i)	Zone number (k)					
	1	2	3	4	5	6
1	0	0	v_1	Kv_1	u_1^a	0
2	v_2	$\alpha_2\left(\dfrac{v_1}{\alpha_1} + \dfrac{w_2}{\alpha_2}\right)$	w_2	Kw_2	u_2^a	v_2
Jump velocity for leading edge of kth zone (D_k)	0	$\dfrac{\alpha_1}{v_1 + w_2}$	0	$\dfrac{\alpha_1\alpha_2^2}{v_2}L^2$	$\dfrac{\alpha_1}{v_2}$	0

Let us assume that the mobilities, the molar charge, and the conductivities depend only on the acidity of the solution, i.e., on the concentration a_1:

$$e_i = e_i(a_1), \quad \gamma_i = \gamma_i(a_1), \quad \sigma_i = \sigma_i(a_1). \tag{2.5}$$

Let us also assume that

$$e_i(a_1) \neq 0. \tag{2.6}$$

This means that the isoelectric points of the components do not belong to the interval of variation in the pH value of the mixture. Let us make the substitutions

$$-e_i(a_1)\,\xi_i = v_i, \quad \xi_i = \frac{-v_i}{e_i(a_1)}, \quad p = \sum_{i=1}^{n} v_i = a_1. \tag{2.7}$$

Then system (2.2)-(2.4) takes on the form

$$\frac{\partial}{\partial t}\left\{\frac{v_i}{e_i(p)}\right\} - \frac{\partial}{\partial x}\left\{\frac{j\,\dfrac{\gamma_i(p)}{e_i(p)}\,v_i}{\displaystyle\sum_{k=1}^{n}\left(1 - \dfrac{\sigma_k(p)}{e_k(p)}\right)v_k}\right\} = 0, \quad i = 1, \ldots, n, \tag{2.8}$$

$$E = \frac{-j}{\sum\limits_{k=1}^{n} \left(1 - \frac{\sigma_k(p)}{e_k(p)}\right) v_k} \cdot \tag{2.9}$$

Let us assume that the concentrations and the current flowing in the mixture are small:

$$v_i = \bar{v}_i \varepsilon, \quad j = \bar{j} \varepsilon, \quad p = \bar{p} \varepsilon, \quad \bar{v}_i = 0(1), \quad \bar{j} = 0(1). \tag{2.10}$$

Here, ε is a small parameter connected with the choice of the magnitude of the current. Let

$$\varepsilon \to 0. \tag{2.11}$$

Then, restricting ourselves to quantities on the order of $O(\varepsilon)$ and omitting the bars above the corresponding quantities, we obtain from (2.8)

$$\frac{\partial}{\partial t}\left\{\frac{v_i}{e_i(0)}\right\} - \frac{\partial}{\partial x}\left\{\frac{\frac{\gamma_i(0)}{e_i(0)} v_i j}{\sum\limits_{k=1}^{n}\left[1 - \frac{\sigma_k(0)}{e_k(0)}\right] v_k}\right\} = 0. \tag{2.12}$$

Here,

$$\gamma_i(0) = \lim_{\varepsilon \bar{p} \to 0} \gamma_i, \quad e_i(0) = \lim_{\varepsilon \bar{p} \to 0} e_i, \quad \sigma_k(0) = \lim_{\varepsilon \bar{p} \to 0} \sigma_k. \tag{2.13}$$

Making the substitution

$$\left[1 - \frac{\sigma_k(0)}{e_k(0)}\right] v_k = u_k, \quad -j\gamma_i(0) = \alpha_i, \tag{2.14}$$

we obtain a system of equations which coincides with (IV.2.1):

$$\frac{\partial u_i}{\partial t} + \frac{\partial}{\partial x}\left(\frac{\alpha_i u_i}{\sum\limits_{k=1}^{n} u_k}\right) = 0, \quad i = 1, \ldots, n. \tag{2.15}$$

As an example, let us consider the case described in Section 4 of Chapter III [see (III.4.11), (III.4.12)],

$$\left.\begin{array}{l}
\sigma_1 = D\dfrac{\text{ch}(\psi - \psi^i)}{\text{ch}(\psi - \psi^e) + m}, \quad e_1 = \dfrac{\text{sh}(\psi - \psi^e)}{\text{ch}(\psi - \psi^e) + m}, \quad \gamma_1 = D\dfrac{\text{sh}(\psi - \psi^i)}{\text{ch}(\psi - \psi^e) + m}, \\[3mm]
\sigma_2 = \dfrac{D_3\alpha}{\alpha + e^\psi}, \quad e_2 = -\dfrac{\alpha}{\alpha + e^\psi}, \quad \gamma_2 = -\dfrac{D_3\alpha}{\alpha + e^\psi}, \\[3mm]
\psi = \ln(-p) = \ln(-\varepsilon\bar{p}), \quad \bar{p} < 0.
\end{array}\right\} \tag{2.16}$$

Then, when $\bar{\varepsilon}p \to 0$, we obtain

$$\sigma_1 = D, \quad e_1 = -1, \quad \gamma_1 = -D,$$

$$\sigma_2 = D_3; \quad e_2 = -1, \quad \gamma_2 = -D_3. \tag{2.17}$$

Let us study the reaction of one moving sample zone with a one-component buffer. In this case, system (2.15) may be rewritten in Riemann invariants [compare with (IV.3.1)-(IV.3.5)]:

$$\frac{\partial R_1}{\partial t} = 0, \tag{2.18}$$

$$\frac{\partial R}{\partial t} + \xi \frac{\partial R}{\partial x} = 0, \tag{2.19}$$

where

$$R_1 = \frac{u_1}{\alpha_1} + \frac{u_2}{\alpha_2}, \quad R = \frac{\alpha_1 u_2 + \alpha_2 u_1}{s}, \quad s = u_1 + u_2,$$

$$\xi = \frac{\alpha_1 u_2 + \alpha_2 u_1}{s^2} = \frac{R}{s} = \frac{\alpha_1 \alpha_2 R_1}{s^2}. \tag{2.20}$$

Let us give the initial conditions for (2.15) when $n = 2$ in the form

$$u_1 = \begin{cases} 0, & x < a_1, \\ v_1, & a_1 < x < a_2, \\ 0, & x > a_2, \end{cases} \quad u_2 = \begin{cases} v_2, & x < a_1, \\ w_2, & a_1 < x < a_2, \\ v_2, & x > a_2, \end{cases} \tag{2.21}$$

where $v_1 = \text{const}$ is the sample concentration in the zone located between the points a_1 and a_2; v_2 is the buffer concentration outside the zone; w_2 is the buffer concentration in the zone occupied by the sample; $(a_2 - a_1)$ is the zone width. Such initial conditions correspond to the case where the sample is located at a specified position in the buffer.

Let us assume that the velocity of the sample zone is greater than the velocity of the buffer, and the velocities are in the same direction:

$$\alpha_1 > \alpha_2 > 0. \tag{2.22}$$

Let us investigate the zone motion using the methods presented in Sections 4-6 of Chapter IV. The initial zone distribution is shown in Fig.

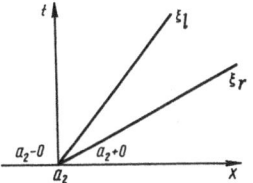

Fig. 28. The limiting characteristic on the (x, t) plane.

26. Zones 1 and 3 correspond to the pure buffer; zone 2 is mixed, and consists of sample and buffer.

It turns out that, in contrast to the cases considered in Sections 4-6 of Chapter IV, the generalized solution to system (2.15) when $t > 0$ is not piecewise-constant: at the boundary between zones 2 and 3, a rarefaction wave appears which is described by a progressive solution (see [40]). Let us find the progressive solution to system (2.15) when $n = 2$ [or equivalently, the solution to system (2.18), (2.19)]. Let us introduce the progressive variable

$$z = \frac{x - x_0}{t - t_0}, \tag{2.23}$$

where x_0, t_0 represent the origin for measuring time and space coordinates. We will look for a solution to (2.18), (2.19) in the form

$$R_1 = R_1(z), \quad R = R(z). \tag{2.24}$$

Considering that

$$(\ )_t = -\frac{x - x_0}{(t - t_0)^2} \cdot (\ )' = -\frac{z}{t - t_0} \cdot (\ )', \quad (\ )_x = \frac{1}{t - t_0}(\ )',$$
$$(\ )' \equiv \frac{\partial}{\partial z}(\ ), \tag{2.25}$$

we rewrite (2.18), (2.19) in the form

$$R_1 = 0, \quad R' \cdot (z - \xi) = 0. \tag{2.26}$$

From this we obtain

$$R_1 = \text{const}, \quad \xi = z. \tag{2.27}$$

The solution $R' = 0$ corresponds to the generalized piecewise-constant solution and does not interest us at present.

From (2.20) using (2.27), we derive

$$R = R(z) = \sqrt{\alpha_1 \alpha_2 R_1 z}\,, \quad R_1 = \text{const.} \tag{2.28}$$

Solving this system [see (2.20)]

$$\frac{u_1(z)}{\alpha_1} + \frac{u_2(z)}{\alpha_2} = R_1 = \text{const},$$

$$\tag{2.29}$$

$$u_1(z) + u_2(z) = \frac{\alpha_1 \alpha_2 R_1}{R(z)}\,,$$

we obtain the progressive solution to (2.15) when $n = 2$:

$$u_1^a(z) = \frac{\alpha_1 \alpha_2 R_1}{\alpha_1 - \alpha_2}\left(\frac{\alpha_1}{R} - 1\right), \quad s^a = u_1^a + u_2^a = \frac{\alpha_1 \alpha_2 R_1}{R}\,,$$

$$u_2^a(z) = \frac{\alpha_1 \alpha_2 R_1}{\alpha_1 - \alpha_2}\left(1 - \frac{\alpha_2}{R}\right). \tag{2.30}$$

From the condition that u_1^a, u_2^a are positive and taking into account (2.22) and (2.28), we obtain

$$0 < \alpha_2 < R(z) < \alpha_1, \quad \frac{\alpha_2}{\alpha_1 R_1} < z < \frac{\alpha_1}{\alpha_2 R_1}\,. \tag{2.31}$$

From the initial state (Fig. 26), zones develop which are represented in Fig. 27. Table 15 corresponds to this figure. The moving jump D_2 starts from the fixed jump D_1. Zone 2 is pure. Zone 3 (Fig. 27) corresponds to zone 2 in Fig. 26. By virtue of the hyperbolicity of system (2.15), the disturbance is propagated with finite velocity; consequently, the concentrations in zone 3 (Fig. 27) will remain as before as long as the jump D_2 does not reach the fixed jump D_3. The two moving jumps D_4, D_5 start from the fixed jump D_3. Zone 4 is mixed, with piecewise-constant concentrations u_1, u_2. Zone 5 corresponds to a rarefaction wave and the solution in this zone has the progressive form. Finally, zone 6 is pure by virtue of the hyperbolicity of the system of equations.

Let us describe the filling in of Table 15. We note the velocities of the fixed jump in columns 1 and 3. The concentrations in columns 1, 3, and 6 are transferred from the initial conditions. Let us fill in column 2. Let us determine the concentration u_1^2 from the condition on the fixed jump D_1:

$$\alpha_1\left[\frac{u_1}{s}\right] = 0, \quad \frac{u_1^1}{s_1} - \frac{u_1^2}{s_2} = 0, \quad u_1^1 \equiv 0, \quad u_1^2 = 0. \tag{2.32}$$

We determine the concentration u_2^2 from the condition that the Riemann invariant R_1 is constant at the moving jump D_2:

$$R_1^2 = R_1^3, \quad \frac{u_2^2}{\alpha_2} = \frac{v_1}{\alpha_1} + \frac{w_2}{\alpha_2}, \quad u_2^2 = \alpha_2 \left(\frac{v_1}{\alpha_1} + \frac{w_2}{\alpha_2} \right). \qquad (2.33)$$

The velocity of the jump D_2 is determined from the Hugoniot conditions for the concentration u_1 at the moving jump:

$$D_2 u_1^3 = \alpha_1 \frac{u_1^3}{s_3}, \quad D_2 = \frac{\alpha_1}{s_3} = \frac{\alpha_1}{v_1 + w_2}. \qquad (2.34)$$

Let us go on to filling in column 4. From the conditions on the fixed jump D_3 for the concentration u_1, we derive the equation

$$\alpha_1 \left(\frac{u_1^4}{s_4} - \frac{v_1}{s_3} \right) = 0, \quad s_3 = v_1 + w_2, \quad s_4 = u_1^4 + u_2^4. \qquad (2.35)$$

We obtain the second equation from the condition that the invariant R_1 is conserved upon going through the moving jumps D_4 and D_5:

$$R_1^4 = R_1^5 = R_1^6, \quad \frac{u_1^4}{\alpha_1} + \frac{u_2^4}{\alpha_2} = \frac{v_2}{\alpha_2}. \qquad (2.36)$$

Solving (2.35) and (2.36) simultaneously, we have

$$u_1^4 = K v_1, \quad u_2^4 = K w_2, \quad K = \frac{\dfrac{v_2}{\alpha_2}}{\dfrac{w_2}{\alpha_2} + \dfrac{v_1}{\alpha_1}}. \qquad (2.37)$$

Before filling in column 5 and determining the velocity of the discontinuity D_4, let us convince ourselves that zone 5 is a rarefaction wave. Let us consider the characteristics coming from the points $(a_2 - 0, 0)$ and $(a_2 + 0, 0)$ on the (x, t) plane [see (2.20) and Fig. 28]:

$$\frac{dx}{dt} \equiv \xi_l \equiv \xi(R_1(a_2 - 0), R(a_2 - 0)) \equiv \frac{\alpha_1 \alpha_2 R_1(a_2 - 0)}{s^2(a_2 - 0)},$$
$$\qquad (2.38)$$

$$\frac{dx}{dt} \equiv \xi_r \equiv \xi(R_1(a_2 + 0), R(a_2 + 0)) \equiv \frac{\alpha_1 \alpha_2 R_1(a_2 + 0)}{s^2(a_2 + 0)}.$$

Using the results obtained (see columns 4 and 6), we derive

$$\xi_l = \frac{\alpha_1 \alpha_2 K \left(\dfrac{v_1}{\alpha_1} + \dfrac{w_2}{\alpha_2} \right)}{K^2 (w_2 + v_1)^2} = \frac{\alpha_1 \alpha_2^2 \left(\dfrac{v_1}{\alpha_1} + \dfrac{w_2}{\alpha_2} \right)^2}{v_2 \left(w_2 + v_1 \right)^2},$$

$$\xi_r = \frac{\alpha_1 \alpha_2 \dfrac{v_2}{\alpha_2}}{v_2^2} = \frac{\alpha_1}{v_2}.$$

(2.39)

Using (2.22), it is easy to show that the stability condition is not satisfied, i.e.,

$$\xi_r > \xi_l.$$
(2.40)

This means that disturbances coming from the point $(a_2 - 0, 0)$ lag behind disturbances coming from the point $(a_2 + 0, 0)$; a rarefaction wave is formed, the concentrations u_1 and u_2 in which are described by the progressive solution (2.30).

Thus, in column 5 we should make the replacements

$$u_1^5 = u_1^a, \quad u_2^5 = u_2^a.$$
(2.41)

The velocities of the leading edge and the trailing edge of the rarefaction wave are determined (see, for example, [40]) by the expression (i.e., by the limiting values of ξ in the rarefaction wave)

$$D_4 = \frac{\alpha_1 \alpha_2^2}{v_2} L^2, \quad D_5 = \frac{\alpha_1}{v_2}, \quad L = \frac{\dfrac{v_1}{\alpha_1} + \dfrac{w_2}{\alpha_2}}{v_1 + w_2}.$$
(2.42)

Thus we have completed Table 15.

We note that the quantity D_5 may be obtained from the Hugoniot conditions for the concentration u_1 at the moving jump D_5. Using (2.28), (2.30), and the conservation conditions for R_1 at D_5, we obtain [comparing with (2.42), we determine z at the right-hand jump]

$$D_5 = \frac{\alpha_1}{s_5} = \frac{\alpha_1}{s^a} = \frac{\alpha_1 R}{\alpha_1 \alpha_2 R_1^6} = \sqrt{z \frac{\alpha_1}{v_2}}.$$
(2.43)

The pattern of zone movement plotted in Fig. 27 and represented in Table 15 is retained right up to the instant of time $t = t_{23}$, i.e., until the moving jump D_2 reaches the fixed jump D_3:

$$t_{23} = \frac{a_2 - a_1}{D_2} = (a_2 - a_1) \cdot \frac{w_2 + v_1}{\alpha_1}.$$
(2.44)

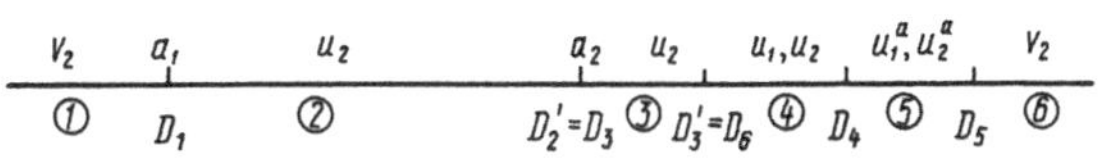

$$V_2 \quad a_1 \quad\quad u_2 \quad\quad\quad a_2 \quad u_2 \quad u_1,u_2 \quad u_1^a,u_2^a \quad V_2$$

$$①\quad D_1 \quad ② \quad\quad D_2'=D_3 \;③\; D_3'=D_6 \;④\; D_4 \;⑤\; D_5 \;⑥$$

Fig. 29. Arrangement of zones in the time interval $t_{23} < t < t_{64}$.

TABLE 16. Concentrations of Substances in the Zones and Jump Velocities Corresponding to Fig. 29

Index of substance (i)	Zone number (k)					
	1	2	3	4	5	6
1	0	0	0	Kv_1	u_1^a	0
2	v_2	$\alpha_2\left(\dfrac{v_1}{\alpha_1}+\dfrac{w_2}{\alpha_2}\right)$	v_2	Kw_2	u_2^a	v_2
Jump velocity for leading edge of kth zone (D_k)	0	$D_2'=0$	$D_3'=D_6=$ $=\dfrac{\alpha_1}{K(v_1+w_2)}$	$\dfrac{\alpha_2\alpha_2^2}{v_2}L^2$	$\dfrac{\alpha_1}{v_2}$	0

For the times $t > t_{23}$, the rearrangement of zones represented in Fig. 29 and Table 16 begins. Columns 1, 4, 5, and 6 in Table 16 coincide with columns 1, 2, 4, and 5 in Table 15. Column 2 in Table 16 is the same as column 2 in Table 15, except for the D row: the zone boundary is the fixed jump $D_3 = D_2' = 0$. Let us describe the filling in of column 3. From the condition for u_1 at the fixed jump D_2', we derive

$$\frac{u_1^2}{s_2} - \frac{u_1^3}{s_3} = 0, \quad u_1^3 = 0. \tag{2.45}$$

We obtain the quantity u_2^3 using the property of conservation of R_1 upon going through the moving jump D_3':

$$\frac{u_1^3}{\alpha_1} + \frac{u_2^3}{\alpha_2} = R_1^4 = K\left(\frac{v_1}{\alpha_1} + \frac{w_2}{\alpha_2}\right), \quad u_2^3 = v_2. \tag{2.46}$$

We determine the velocity D_3' from the condition for u_1 at the moving jump D_3':

$$D_3' \equiv D_6 = \frac{\alpha_1}{s_4} = \frac{\alpha_1}{K(v_1 + w_2)}. \tag{2.47}$$

Thus we have filled in Table 16.

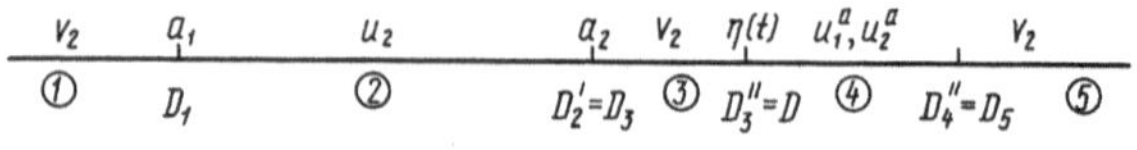

$$V_2 \quad q_1 \qquad u_2 \qquad\qquad q_2 \; V_2 \; \eta(t) \; u_1^a,u_2^a \quad V_2$$
$$\textcircled{1} \quad D_1 \qquad \textcircled{2} \qquad\quad D_2'=D_3 \;\textcircled{3}\; D_3''=D \;\textcircled{4}\; D_4''=D_5 \;\textcircled{5}$$

Fig. 30. Final arrangement of zones when $t > t_{64}$.

TABLE 17. Concentrations of Substances in the Zones and Jump Velocities Corresponding to Fig. 30

Index of substance (i)	Zone number (k)				
	1	2	3	4	5
1	0	0	0	u_1^a	0
2	v_2	$\alpha_2\left(\dfrac{v_1}{\alpha_1}+\dfrac{w_2}{\alpha_2}\right)$	v_2	u_2^a	v_2
Jump velocity for leading edge of kth zone (D_k)	0	0	$\dfrac{\alpha_1}{v_2}\left(1-\dfrac{c}{\sqrt{t}}\right)$	$\dfrac{\alpha_1}{v_2}$	—

It is interesting to note that, *independently* of later zone rearrangements, the buffer solution has "*remembered*" the information about the location of the sample zone and its concentration:

$$u_2(x,\,t) = w_2 + \frac{\alpha_2}{\alpha_1}v_1, \quad t>t_{23}, \quad a_1<x<a_2. \tag{2.48}$$

This information may be "*erased*" as a result of diffusion *only* after the electric current *stops* passing through the solution.

It is easy to show that the jump velocity D_3' is higher than the jump velocity D_4:

$$D_3' \equiv D_6 = \frac{\alpha_1}{K\,(v_1+w_2)} > \frac{\alpha_2^2\alpha_1}{v_2}L^2 = D_4. \tag{2.49}$$

Consequently, at the instant of time $t = t_{64}$, when the jump D_3' reaches the jump D_4, one more zone rearrangement occurs (this time, the final zone rearrangement):

$$t_{64} = \frac{D_4 l_{23}}{D_6-D_4} + t_{23} = \frac{D_6}{D_6-D_4}\,t_{23},$$
$$t_{64} = (a_2 - a_1)\frac{(v_1+w_2)^2}{(\alpha_1-\alpha_2)\,v_1}. \tag{2.50}$$

The final zone distribution is shown in Fig. 30 and in Table 17. The procedure for filling in this table, except for the D row in columns 2 and 3,

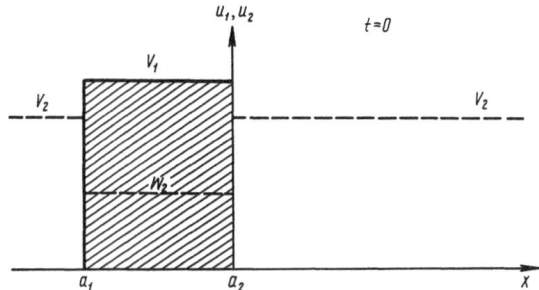

Fig. 31. Position of zones at the initial instant of time $t = 0$.

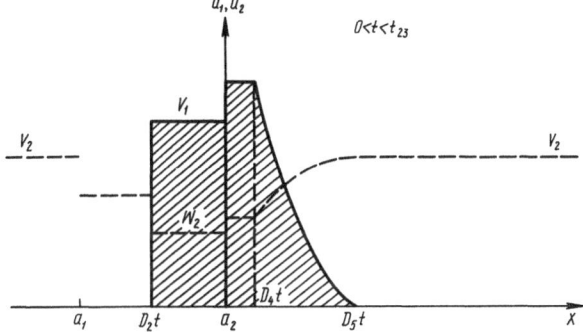

Fig. 32. Zone evolution pattern in the time interval $0 < t < t_{23}$.

is analogous to the procedure for filling in Table 16: columns 1, 2, 3, 4, and 5 in Table 17 coincide with columns 1, 2, 3, 5, and 6 in Table 16.

Let us determine the velocity of the trailing edge of the rarefaction wave passing through the fixed jump. Let us designate as $\eta(t)$ the boundary between zones 3 and 4 (see Fig. 30). As the origin, let us choose $x_0 = a_2$ [for convenience, we set $a_2 = 0$ ($t_0 = 0$)]. Then [see (2.23)], for the progressive variable we obtain

$$z = \frac{x}{t} = \frac{\eta(t)}{t}. \tag{2.51}$$

From the condition for the concentration u_1 at the moving jump $D_3'' \equiv D$, we derive

$$D = \frac{\alpha_1}{s_4} = \frac{\alpha_1}{s^a} = \frac{\alpha_1 R}{\alpha_1 \alpha_2 R_1^4} = \frac{R}{v_2} = \frac{\sqrt{\alpha_1 \alpha_2 R_1^4 z}}{v_2} = \sqrt{\frac{\alpha_1}{v_2} z}. \tag{2.52}$$

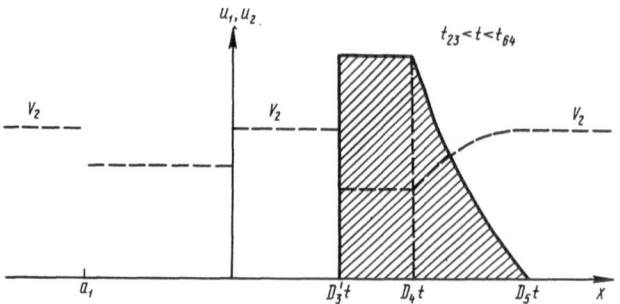

Fig. 33. Schematic zone profile in the time interval $t_{23} < t < t_{64}$.

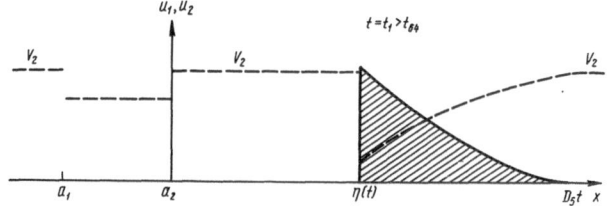

Fig. 34. Zone evolution pattern after the instant of time $t = t_{64}$. The function $\eta(t)$ is represented in (2.56).

Taking into account the fact that the velocity D is specified by the expression

$$D = \frac{d\eta(t)}{dt}, \tag{2.53}$$

we obtain the equation for the definition of $\eta(t)$:

$$\frac{d\eta(t)}{dt} = \sqrt{D_5 \frac{\eta(t)}{t}} \tag{2.54}$$

with initial condition

$$\eta(t_{64}) = D_4 t_{64}. \tag{2.55}$$

This condition corresponds to the fact that, at the instant of time $t = t_{64}$, when the jump D_3' reaches the jump D_4 (see Fig. 29), rearrangement of the zones begins, as a result of which we see a change in the velocity of the interface between the formed zones 3 and 4 in Fig. 30.

Solving (2.54), we obtain

$$\eta(t) = D_5 t - 2D_5 C \sqrt{t} + D_5 C^2, \tag{2.56}$$

$$D = \frac{d\eta(t)}{dt} = D_5\left(1 - \frac{C}{\sqrt{t}}\right). \tag{2.57}$$

We determine the constant C from the condition (2.55), taking into account (2.42) and (2.50):

$$C = \sqrt{t_{64}}\left(1 - \sqrt{\frac{D_4}{D_5}}\right) = \sqrt{\frac{(\alpha_1 - \alpha_2)(a_2 - a_1)v_1}{\alpha_1^2}} > 0. \tag{2.58}$$

It is easy to calculate the width of the interval $\delta(t)$ in which the sample u_1 is concentrated:

$$\begin{aligned} \delta(t) &= a_2 + D_5 t - (a_1 + D_2 t), \quad t < t_{23}, \\ \delta(t) &= D_5 t - D_3(t - t_{23}), \quad t_{23} < t < t_{64}, \\ \delta(t) &= D_5 t - \eta(t), \quad t > t_{64}. \end{aligned} \tag{2.59}$$

Using the tables, we obtain

$$\delta(t) = \begin{cases} (a_2 - a_1) + \dfrac{\alpha_1}{v_2}\left(\dfrac{(v_1 - v_2 + w_2)}{v_1 + w_2}\right)t, & 0 \leqslant t < t_{23}, \\[2mm] (a_2 - a_1)\dfrac{w_2 + v_1}{v_2} + \dfrac{\alpha_1 - \alpha_2}{\alpha_2} \cdot \dfrac{v_1}{w_2 + v_1}(t - t_{23}), & t_{23} < t < t_{64}, \\[2mm] \dfrac{C\alpha_1}{v_2}(2\sqrt{t} - G), & t > t_{64}. \end{cases} \tag{2.60}$$

We note that, beginning at the instant of time $t = t_{64}$, the total concentration M of the sample u_1, concentrated in the interval $(\eta(t), D_5 t)$, is equal to the initial total concentration M_1:

$$M = \int_{\eta(t)}^{D_5 t} u_1^a dx = M_1, \quad M_1 \equiv v_1(a_2 - a_1). \tag{2.61}$$

From formula (2.60), it is quite evident that over the course of time $(t > t_{64})$, the length of the sample zone becomes "almost infinite," i.e., on the order of magnitude $O(\sqrt{t})$, $t \to \infty$. In this case, the sample concentration in the zone is "almost equal to zero," i.e., it becomes on the order of $O(t_{-1/2})$, $t \to \infty$.

It is interesting to note that the ratio of the sample concentration to the buffer concentration everywhere the solution is piecewise-constant and different from zero is equal to its initial value (see Tables 15 and 16):

$$\frac{u_1}{u_2} = \frac{v_1}{w_2}, \quad t < t_{64}. \tag{2.62}$$

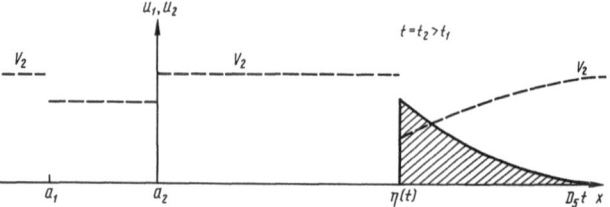

Fig. 35. Later zone evolution after the instant of time $t = t_{64}$.

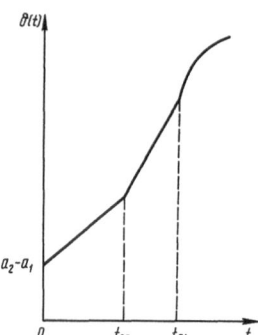

Fig. 36. Variation in the width of the sample zone over time.

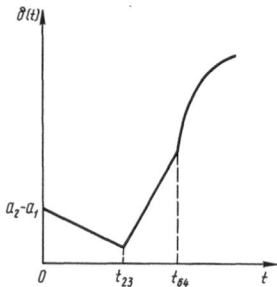

Fig. 37. Variation in the width of the sample zone over time when the relation $v_2 > v_1 + w_2$ is satisfied.

The latter means that even if in any section of the electrophoretic column we can increase the sample concentration, i.e., make $u_1 > v_1$ (this is possible in the case when $v_2 > v_1 + w_2$; see Tables 15 and 16), the buffer concentration also increases.

In Figs. 31-35 we show the evolution of the zone profile for the case under investigation. In Figs. 36 and 37 we present the time dependence for the width of the interval occupied by the sample u_1. Figure 37 corresponds to the case where the following relation is satisfied:

$$v_2 > v_1 + w_2. \tag{2.63}$$

The results obtained require experimental verification, which apparently is not difficult to do on any instrument designed for performing isotachophoresis. As far as the authors know, such experiments have not been carried out.

3. Radial Distortion of the Zone Profile

In this section, we consider the simplest non-one-dimensional model for zone electrophoresis for the case where the sample mobility depends on the temperature. In this model, the motion of the sample zone does not cause changes in such basic characteristics of the mixture, as a whole, as acidity and electrical conductivity; consequently, the temperature of the mixture also does not change. The difference between this case and the model in Section 1 of this chapter involves the fact that, for a known distribution of the temperature field in the electrophoretic column, the sample mobility is a specified function of x and t. Taking into account the temperature dependence of the mobility allows us to describe the distortion of the concentration profile of the sample zone in the electrophoretic column in the radial direction.

Let us consider an electrophoretic column of infinite length with inner radius r_0. For the axially symmetric case, the system of equations describing the motion of the zone has the form [see (1.1), (1.13), (III.5.16)]

$$\frac{\partial \xi}{\partial t} - \mu_0 \frac{\partial^2 \xi}{\partial z^2} - \mu_0 \frac{1}{r} \frac{\partial}{\partial r} r \frac{\partial \xi}{\partial r} + \gamma(T) E \frac{\partial \xi}{\partial z} = 0,$$

$$x = (z, r, 0), \quad 0 \leqslant r \leqslant r_0, \quad -\infty < z < \infty \tag{3.1}$$

with initial condition

$$\xi(r, z, t)|_{t=0} = \xi_0(z) \tag{3.2}$$

and boundary conditions

$$\frac{\partial \xi(z, r, t)}{\partial r}\bigg|_{r=r_0} = 0, \quad \xi(z, r, t)|_{z=\mp\infty} = 0, \tag{3.3}$$

where z is the distance along the axis of the cylinder, r is the instantaneous radius, $\gamma(T)$ is the sample mobility (dependent on temperature), and $\xi_0(z)$ is the initial concentration distribution of the sample (independent of r).

In order to determine the temperature in the column, we have the thermal conductivity equation

$$\frac{\partial T}{\partial t} - \chi \Delta T = \delta Ej = \text{const} > 0 \qquad (3.4)$$

with initial condition

$$T(z, r, t)|_{t=0} = T_0 = \text{const} \qquad (3.5)$$

and boundary conditions

$$T(z, r, t)|_{r=r_0} = T_0 = \text{const}, \qquad (3.6)$$

$$T(z, r, t)|_{z=\mp\infty} = T_0 = \text{const}. \qquad (3.7)$$

Condition (3.5) corresponds to the case where the temperature of the mixture at the initial instant of time is constant. Conditions (3.6) and (3.7) mean that the specified constant temperature is maintained on the lateral surfaces and the ends of the column. Solving the problem (3.4)-(3.7), we determine the temperature distribution in the column. Let us assume that the temperature along the column does not change, i.e.,

$$T(r, z, t) = T(r, t). \qquad (3.8)$$

We will look for a solution to (3.4)-(3.6) in the form

$$T(r, t) = u(r, t) + T_0 - \frac{\delta_0}{4\chi}(r^2 - r_0^2), \quad \delta_0 \equiv \delta Ej > 0. \qquad (3.9)$$

Then, in order to determine $u(r, t)$, we have the problem

$$\frac{\partial u}{\partial t} - \chi \frac{1}{r} \frac{\partial}{\partial r} r \frac{\partial u}{\partial r} = 0, \qquad (3.10)$$

$$u|_{r=r_0} = 0, \quad u|_{t=0} = \frac{\delta_0}{4\chi}(r^2 - r_0^2), \quad u|_{r=0} < \infty. \qquad (3.11)$$

Using the separation of variables (see, for example [24, p. 475], where the solution to the problem is presented), we obtain

$$u(r, t) = \frac{2}{r_0^2} \sum_{n=1}^{\infty} \frac{J_0\left(\frac{\lambda_n r}{r_0}\right)}{J_1^2(\lambda_n)} e^{-\chi\left(\frac{\lambda_n}{r_0}\right)^2 t} \cdot \int_0^{r_0} r \left\{ -\frac{\delta_0}{4\chi}(r^2 - r_0^2) J_0\left(\frac{\lambda_n r}{r_0}\right)\right\} dr,$$
$$(3.12)$$

where J_0 and J_1 are Bessel functions, and λ_n are the roots to the function J_0. Let us apply familiar formulas (see, for example, [4])

$$\int_0^{\lambda_n} z^k J_0(z)\, dz = \lambda_n J_1(\lambda_n)\, S_{k,0}(\lambda_n), \tag{3.13}$$

$$S_{3,0}(z) = z^2 - 4S_{1,0}(z), \quad S_{1,0}(z) = 1, \tag{3.14}$$

where $S_{3,0}$ and $S_{1,0}$ are Lommel functions. Then the final solution has the form

$$T(r, t) = T_0 - \frac{\delta_0}{4\chi}(r^2 - r_0^2) + \frac{\delta_0 r_0^2}{2\chi} \sum_{n=1}^{\infty} \frac{J_0\left(\frac{\lambda_n r}{r_0}\right)}{J_1(\lambda_n)} e^{-\chi\left(\frac{\lambda_n}{r_\bullet}\right)^2 t}$$

$$\times \left(\frac{1}{\lambda_n} - 1 - \frac{4}{\lambda_n^3}\right). \tag{3.15}$$

Then let us consider the case where the temperature distribution in the column is steady state and we neglect sample diffusion:

$$T(r, t) = T_0 - \frac{\delta_0}{4\chi}(r^2 - r_0^2), \tag{3.16}$$

$$\mu_0 = 0. \tag{3.17}$$

In order to determine the concentration $\xi(z, r, t)$, we have

$$\frac{\partial \xi}{\partial t} + \gamma(T) E \frac{\partial \xi}{\partial z} = 0. \tag{3.18}$$

The solution to this equation, taking into account the initial condition (3.2), has the form

$$\xi(z, r, t) = \xi_0(z - \omega(r) \cdot t), \tag{3.19}$$

$$\omega(r) \equiv \gamma\left(T_0 - \frac{\delta_0}{4\chi}(r^2 - r_0^2)\right). \tag{3.20}$$

Thus, the motion of the sample zone occurs at the velocity $\omega(r)$, depending on the distance from the cylinder axis.

Usually the mobility increases with an increase in temperature:

$$\frac{\partial \gamma(T)}{\partial T} > 0. \tag{3.21}$$

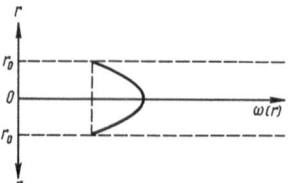

Fig. 38. Profile of the
sample motion as a function
of the distance from the axis
of the electrophoretic column.

In this case, the velocity of the zone will decrease with an increase in the
distance from the axis of the electrophoretic column (Fig. 38), which leads
to distortion of the zone profile.

Chapter VI

CREATION OF A pH GRADIENT
IN INFINITE-COMPONENT SYSTEMS

As has been indicated in Chapter III, in order to carry out the isoelectric focusing method we need to have a monotonic pH profile in the
specified range. In practice, the required pH gradient may be created using
buffer solutions containing a relatively small number of components. If in
this case we do not require a special initial distribution of the solution components in the electrophoretic chamber and the gradient is formed under the
influence of an electric field during electrophoresis, then such a pH gradient is called "natural." Usually, in order to obtain a high-quality pH gradient, we require such a large number of components that such a mixture is
more conveniently considered as an infinite-component mixture (see
Chapter VII), although examples are known when natural pH gradients are
created using a small number of components: mixture of several buffers
containing several amino acids [66, 96].

Another widely used method for creating pH gradients involves so-
called artificial pH gradients, specified by the initial concentration distribution of the components of the buffer solution in the electrophoretic chamber. A serious obstacle to accomplishing this method is the insufficient
stability of artificial pH gradients: the initial pH distribution varies during
the electrophoresis process. In this case, disappearance of the pH gradient
is possible (the pH distribution becomes constant along the chamber), or
motion (drift) of the pH gradient as a whole may occur in the direction of
one of the electrodes. Thus, for example, motion of pure zones in isotachophoresis may be used to create a stable pH gradient, moving as a whole
and located between the terminator and the leader (Chapter IV, Section 4).

From a practical standpoint, apparently one of the simplest sets of
components for creating an artificial pH gradient is boric acid and some
polyhydroxy compounds (for example, glycerine). The effectiveness of
such systems was first demonstrated in the work of Troitskii et al. [49-51],

who proposed the given method. The authors presented the chemical and analytical principles of interest and original methodological approaches, and also developed an apparatus for carrying out isoelectric focusing in borate–polyol systems.

Later in this chapter we construct a mathematical model which allows us to describe the creation and evolution of an artificial pH gradient in a multicomponent boric acid–polyol system. Such electrolytes are weak, which allows us to make important simplifications. The system of equations with boundary and initial conditions for studying the process of formation of a pH gradient is given in Sections 2-4.

The evolution of a continuous initial pH distribution for vanishingly small diffusion is investigated in Sections 5 and 6 for the boric acid–polyol system. We should note that the one-dimensional case considered is not only the simplest case but is also the most frequently realized case in practice (in commonly used electrophoretic columns, the length-to-diameter ratio is sufficiently large). Finally, in Section 7 we obtain the solution for a multicomponent mixture consisting of boric acid and several polyols in the case of a piecewise-constant initial concentration distribution. We give an algorithm which allows us to calculate the evolution of the pH profile in the column. Here, we also indicate the range of variation in the concentrations of the mixture components, which allows us to ensure stability of the pH gradient. It turns out that this condition is satisfied for component concentrations used in practice. We should point out that one advantage of the borate–polyol method is connected with the fact that, upon creation of the initial pH gradient, simultaneously a density gradient for the liquid is formed which is directed downward and consequently protects this system from gravitational convective instability (let us recall that in practice the pH gradient is obtained by stacking portions of polyol with gradually decreasing concentrations in a vertical column). However, such stabilization does not protect the system from electrolytic mixing (see the introduction to Chapter IV).

1. Reactions Occurring in Aqueous Solutions of Boric Acid with Polyols

In aqueous solutions of boric acid, an equilibrium occurs between three-coordinate and four-coordinate boron atoms, which in the general case is strongly shifted to the left. The dissociation reaction proceeds according to the scheme [27, 55]

$$H_3BO_3 + H_2O \rightleftarrows H^+ + B(OH)_4^-.$$
$$(HB \rightleftarrows H^+ + B^-)$$

$$(1.1)$$

TABLE 18. Equilibrium Constants for 1:1 and 1:2 Complexes of Boric Acid with Some Polyols in a 0.1 M KCl Solution at 25°C, $K_0 = 10^{-9.24}$ (from [58])

Polyol	$\lg K_1'$	$\lg (K_1' K_2')$	$-\lg (K_0 K_1')$	$-\lg (K_0 K_1' K_2')$
L-rhamnose	—	2.61	—	6.63
D-galactose	—	2.39	—	6.85
D-glucose	2.02	2.63	7.22	6.61
D-arabinose	—	3.28	—	5.96
D-xylose	—	4.01	—	5.23
D-mannose	—	4.52	—	4.72
D-fructose	3.4	4.9	5.84	4.34
L-sorbose	—	5.8	—	3.44
D-mannitol	4.00	4.88	5.24	4.36
D-sorbitol	—	5.65	—	3.59
D-dulcitol	—	5.23	—	4.01
Glycerine	—	1.35	—	7.89

The existence of three types of compounds of boric acid with polyols has been determined [55]:

1) compounds of the ester type A:

$$-\overset{|}{\underset{|}{C}}-O\diagdown \atop -\overset{|}{\underset{|}{C}}-O\diagup \quad B-OH;$$

2) monodiolborates of the type BD^- (1:1 complex):

$$\left[-\overset{|}{\underset{|}{C}}-O \diagdown \atop -\overset{|}{\underset{|}{C}}-O \diagup B \diagup {OH} \atop \diagdown OH \right]^-;$$

3) Didiolborates of the type BD_2^- (1:2 complex):

$$\left[-\overset{|}{\underset{|}{C}}-O\diagdown \quad \diagup O-\overset{|}{\underset{|}{C}}- \atop -\overset{|}{\underset{|}{C}}-O\diagup B \diagdown O-\overset{|}{\underset{|}{C}}- \right]^-.$$

The reactions of formation for these compounds proceed according to the scheme [55]

$$B(OH)_4^- + \begin{array}{c} HO-\overset{|}{\underset{|}{C}}- \\ HO-\overset{|}{\underset{|}{C}}- \end{array} \rightleftarrows 2H_2O + \begin{array}{c} HO \\ HO \end{array}\!\!>\!\!B\!\!<\!\!\begin{array}{c} O-\overset{|}{\underset{|}{C}}- \\ O-\overset{|}{\underset{|}{C}}- \end{array}$$

$$(B^- + D \rightleftarrows BD^-), \tag{1.2}$$

$$\left[\begin{array}{c} -\overset{|}{\underset{|}{C}}-O \\ -\overset{|}{\underset{|}{C}}-O \end{array}\!\!>\!\!B\!\!<\!\!\begin{array}{c} OH \\ OH \end{array}\right]^- + \begin{array}{c} HO-\overset{|}{\underset{|}{C}}- \\ HO-\overset{|}{\underset{|}{C}}- \end{array} \rightleftarrows 2H_2O +$$

$$+ \left[\begin{array}{c} -\overset{|}{\underset{|}{C}}-O \\ -\overset{|}{\underset{|}{C}}-O \end{array}\!\!>\!\!B\!\!<\!\!\begin{array}{c} O-\overset{|}{\underset{|}{C}}- \\ O-\overset{|}{\underset{|}{C}}- \end{array}\right]^{--}$$

$$(BD^- + D \rightleftarrows BD_2^-), \tag{1.3}$$

$$\left[\begin{array}{c} -\overset{|}{\underset{|}{C}}-O \\ -\overset{|}{\underset{|}{C}}-O \end{array}\!\!>\!\!B\!\!<\!\!\begin{array}{c} OH \\ OH \end{array}\right]^- + H^+ \rightleftarrows \begin{array}{c} -\overset{|}{\underset{|}{C}}-O \\ -\overset{|}{\underset{|}{C}}-O \end{array}\!\!>\!\!B-OH + H_2O$$

$$(BD^- + H^+ \rightleftarrows A). \tag{1.4}$$

The equilibrium constants of these reactions are

$$\frac{[H^+][B^-]}{[HB]} = K_0, \quad \frac{[BD^-]}{[B^-][D]} = K_1', \quad \frac{[BD_2^-]}{[BD^-][D]} = K_2',$$
$$\frac{[A]}{[BD^-][H^+]} = K_3'. \tag{1.5}$$

For a sufficiently high relative polyol concentration, reactions of type (1.4) in practice do not occur, i.e., there are no compounds of the ester type A ($K_3' \equiv 0$). We also note [6, 55] that mainly complexes of the 1:2 type are formed; this is explained by the fact that, in the tetrahedral borate anion, there are no steric hindrances to addition of even large polyol molecules. As an example, let us give the equilibrium constants for 1:1 and 1:2 complexes with some polyols in a 0.1 M KCl solution at 25°C (Table 18), constructed from data given in [58].

Obviously, in order to create a mathematical model for isoelectric focusing in borate–polyol systems which most accurately describes the processes occurring, we need to know quite accurately just how the chemical reactions occur in the system. Unfortunately, as far as we know, at the present time such data are not available. Therefore, in the following development we will restrict ourselves to consideration of such borate–polyol systems in which the polyol concentration is sufficiently high, i.e., in solution the 1:2 complexes are preferentially formed. In this case, the scheme for the reactions occurring in solution has the form [6]

$$HB \rightleftarrows H^+ + B^-, \tag{1.6}$$

$$HB + 2D \rightleftarrows H^+ + BD_2. \tag{1.7}$$

The equilibrium constants for these reactions are determined by the expressions

$$\frac{[H^+] [B^-]}{[HB]} = K_0, \quad \frac{[H^+] [BD_2^-]}{[HB] [D]^2} = K_0 K_1' K_2' \equiv K. \tag{1.8}$$

If the solution contains several polyols of different types, we will assume that there is no direct reaction between pairs of different polyols, and the reaction scheme appears as follows:

$$HB \rightleftarrows H^+ + B^-, \tag{1.9}$$
$$HB + 2D_i \rightleftarrows H^+ + B(D_i)_2^- \quad (i = 1, \ldots, n)$$

with equilibrium constants

$$\frac{[H^+] [B^-]}{[HB]} = K_0, \quad \frac{[H^+] [B(D_i)_2^-]}{[HB] [D_i]^2} = K_i \quad (i = 1, \ldots, n), \tag{1.10}$$

where D_i is the polyol of type i; K_i is the equilibrium constant for the polyol of type i; n is the number of polyols of different types in the solution.

2. "Slow" Variables for Describing Borate–Polyol Systems

Let us consider a $(2n + 4)$-component mixture composed of a solvent (water), H^+ ions, boric acid HB, borate ions B^-, polyols $D_r (r = 1, \ldots, n)$, and dipolyolborate ions $B(D_r)_2^- (r = 1, \ldots, n)$, in which reversible chemical reactions of type (1.9) occur. Based on the results of Chapter II, let us introduce the "slow" variables describing the behavior of such a mixture. Let us write the chemical kinetics equations for the molar concentrations based on scheme (1.9) (in dimensional variables):

$$\frac{\partial a_1}{\partial t} = \sum_{r=1}^{n} \sigma^{(r)} + \sigma^{(n+1)}, \tag{2.1}$$

$$\frac{\partial a_{2r}}{\partial t} = - 2\sigma^{(r)}; \tag{2.2}$$

$$\frac{\partial a_{2r+1}}{\partial t} = \sigma^{(r)}, \tag{2.3}$$

$$\frac{\partial a_{2n+3}}{\partial t} = \sigma^{(n+1)}, \tag{2.4}$$

$$\frac{\partial a_{2n+2}}{\partial t} = - \sum_{r=1}^{n} \sigma^{(r)} - \sigma^{(n+1)}, \tag{2.5}$$

$$\sigma^{(r)} = - k_r^+ (a_{2r})^2 a_{2n+2} + k_r^- a_1 a_{2r+1}, \quad r = 1, \ldots, n, \tag{2.6}$$

$$\sigma^{n+1} = - K_{n+1}^+ a_{2n+2} + K_{n+1}^- a_1 a_{2n+3}. \tag{2.7}$$

Here

$$a_1 = [\text{H}^+], \quad a_{2r} = [D_r],$$
$$a_{2r+1} = [\text{B} (D_r)_2^-], \quad r = 1, \ldots, n; \quad a_{2n+3} = [\text{B}^-], \tag{2.8}$$
$$a_{2n+2} = [\text{HB}].$$

Then the integrals for Eqs. (2.1)-(2.5), which we take as the "slow" variables, have the form

$$\xi_r = a_{2r} + 2a_{2r+1}, \quad r = 1, \ldots, n,$$
$$\xi_{n+1} = \sum_{r=1}^{n} a_{2r+1} + a_{2n+2} + a_{2n+3}, \tag{2.9}$$
$$\xi_0 = a_1 - \sum_{r=1}^{n} a_{2r+1} - a_{2n+3}.$$

In this case, the corresponding sets of indices (see Chapter II) appear as follows:

$$I_r = \{2r,\ 2r+1\}, \quad r = 1,\ \ldots,\ n,$$

$$I_{n+1} = \{2n+2;\ 2n+3;\ 2r+1,\quad r = 1,\ \ldots,\ n\},$$

$$I_0 = \{1;\ 2n+3;\ 2r+1,\quad r = 1,\ \ldots,\ n\},$$

$$I_{0k} = \{2k+1\}, \quad k = 1,\ \ldots,\ n+1. \tag{2.10}$$

Going to dimensionless variables according to (II.3.1), we write the condition for local chemical equilibrium ($\sigma^{(r)} = 0,\ r = 1,\ \ldots,\ n+1$):

$$\frac{a_1 \cdot a_{2r+1}}{a_{2n+2} \cdot (a_{2r})^2} = K_r K_*, \quad r = 1,\ \ldots,\ n, \tag{2.11}$$

$$\frac{a_1 \cdot a_{2n+3}}{a_{2n+2}} = \frac{K_{n+1}}{K_*} \quad (K_{n+1} = K_0), \tag{2.12}$$

$$\dim K_* = \frac{\text{moles}}{\text{m}^3}, \quad \dim K_{n+1} = \frac{\text{moles}}{\text{m}^3}, \quad \dim K_r = \frac{\text{m}^3}{\text{mole}}. \tag{2.13}$$

Relations (2.9), (2.11), and (2.12) allow us to obtain the formulas of the type (II.1.26) and (II.1.27) for going from the variables $a_1,\ \ldots,\ a_{2n+3}$ to the "slow" variables $\xi_0,\ \ldots,\ \xi_{n+1}$. However, direct determination of the functions φ_s ($s = 1,\ \ldots,\ 2n+3$) is a rather complicated problem. Fortunately, in many cases we do not need to solve the problem of determining the φ_s exactly. It is sufficient to restrict ourselves to determining the approximate functional relationships between a_s ($s = 1,\ \ldots,\ 2n+3$) and ξ_k ($k = 0,\ \ldots,\ n+1$).

Let us consider the case where $K_r K_* \to 0$, $K_{n+1}/K_* \to 0$ (see Table 18). Let us set

$$\frac{K_{n+1}}{K_*} = \alpha_*^2, \quad K = \max_{1 \leqslant r \leqslant n} K_r, \quad \beta_r = \frac{K_r}{K} = O(1), \quad r = 1,\ \ldots,\ n, \tag{2.14}$$

$$K_* = \sqrt{\frac{K_{n+1}}{K}}, \quad \alpha_* = (K K_{n+1})^{1/4}, \quad \alpha_* \to 0, \quad n \ll \frac{1}{\alpha_*},$$

$$w = \alpha_* a_1, \quad w = O(1), \quad \alpha_* \to 0. \tag{2.15}$$

Let us determine the degrees of dissociation [see (II.1.27)]

$$a_{2r} = \beta_r^{2r} \xi_r, \quad 2a_{2r+1} = \beta_r^{2r+1} \xi_r, \quad \alpha_{2r+1}^r = 2$$
$$a_{2n+2} = \beta_{n+1}^{2n+2} \xi_{n+1}, \quad a_{2n+3} = \beta_{n+1}^{2n+3} \xi_{n+1},$$
$$a_{2r+1} = \beta_{n+1}^{2r+1} \xi_{n+1},$$
$$(\beta_r^{2r} + \beta_r^{2r+1}) = 1,$$
$$\sum_{r=1}^{n} \beta_{n+1}^{2r+1} + \beta_{n+1}^{2n+2} + \beta_{n+1}^{2n+3} = 1, \quad r = 1, \ldots, n. \tag{2.16}$$

Then from (2.11) and (2.12), using (2.14)-(2.16), we derive

$$\frac{\beta_{n+1}^{2r+1}}{\left(1 - \sum_{r=1}^{n} \beta_{n+1}^{2r+1} - \beta_{n+1}^{2n+3}\right)(1 - \beta_r^{2r+1})^2} = \frac{\alpha_* (\xi_r)^2}{w} \beta_r, \tag{2.17}$$

$$\frac{\beta_{n+1}^{2n+3}}{\left(1 - \sum_{r=1}^{n} \beta_{n+1}^{2r+1} - \beta_{n+1}^{2n+3}\right)} = \frac{\alpha_*}{w}, \tag{2.18}$$

$$\beta_r^{2r+1} = 2\beta_{n+1}^{2r+1} \frac{\xi_{n+1}}{\xi_r} \quad (r = 1, \ldots n). \tag{2.19}$$

It is easy to show, for example, looking for a solution in the form of a series in powers of α_*, that the following relations are valid:

$$\begin{aligned}
\beta_{n+1}^{2r+1} &= \frac{\alpha_* (\xi_r)^2}{w} \beta_r + O(\alpha_*^2), \quad \beta_{n+1}^{2n+3} = \frac{\alpha^*}{w} + O(\alpha_*^2), \\
\beta_r^{2r+1} &= 2\alpha_* \beta_r \frac{\xi_r \cdot \xi_{n+1}}{w} + O(\alpha_*^2), \quad \beta_r^{2r} = 1 + O(\alpha_*) \\
(r &= 1, \ldots n); \quad \beta_{n+1}^{2n+2} = 1 + O(\alpha_*), \quad \alpha_* \to 0.
\end{aligned} \tag{2.20}$$

Using the results obtained, from formula (II.1.31) using (2.10), (2.15), and (2.20), we obtain an expression for the molar charge:

$$\xi_0 = \alpha_* \left\{ w - \frac{\xi_{n+1}}{w} - \sum_{r=1}^{n} \beta_r \frac{(\xi_r)^2 \xi_{n+1}}{w} \right\} + O(\alpha_*^2). \tag{2.21}$$

From this, making the substitution

$$\xi_0 = \xi \cdot \alpha_*, \tag{2.22}$$

we obtain

$$w = \frac{\xi + \sqrt{\xi^2 + 4\xi_{n+1} + 4\sum_{r=1}^{n} \beta_r \, (\xi_r)^2 \, \xi_{n+1}}}{2} + O\,(\alpha_*). \tag{2.23}$$

Thus, the quantity w is also an integral of system (2.1)-(2.5) as a function of the integrals $\xi_0, \xi_1, \ldots, \xi_{n+1}$ and may be chosen as the "slow" variable.

3. Mathematical Model for Creating pH Gradients in Borate–Polyol Systems

From the results of the preceding section, it follows that the description of processes for an original $(2n + 4)$-component mixture is reduced, in the case of local chemical equilibrium, to a description of processes in an $(n + 1)$-component mixture of chemical subsystems. The equations for describing such a mixture have the form (II.3.4)-(II.3.16). Let us make additional simplifying assumptions.

We will assume that the mixture of chemical subsystems as a whole is found in mechanical equilibrium, the density of the mixture is constant, there are no thermal diffusion and barodiffusion effects, the dielectric constant of the mixture is equal to the dielectric constant of the solvent, the reactions occurring are isothermal, and the Einstein relations are satisfied for the diffusion coefficients and the mobilities (in the original mixture):

$$\left. \begin{aligned} &v = 0, \quad \rho = 1, \quad \varepsilon = 1, \quad \gamma_s = z_s D_s, \\ &h_s = h_0, \quad D_s^T \equiv D_s^p \equiv 0, \quad s = 1, \ldots, 2n + 4; \\ &z_{2s+1} = -1, \quad z_{2s} = 0, \quad s = 1, \ldots n + 1; \quad z_1 = 1. \end{aligned} \right\} \tag{3.1}$$

Furthermore, we assume that

$$E = O\left(\frac{1}{\alpha_*}\right), \quad \delta = O\,(\alpha_*), \quad \theta = O\,(\alpha_*^3), \quad r = O\left(\frac{1}{\alpha_*}\right), \quad \alpha_* \to 0. \tag{3.2}$$

We note that the requirement $\theta = O(\alpha_*^3)$ is equivalent to the assumption that the solution is electrically neutral. We make the substitutions

$$\delta = \delta_* \alpha_*, \quad \theta = \theta_0 \alpha_*^3, \quad E = \frac{1}{\alpha_*} E_*, \quad r = \frac{1}{\alpha^*} r_*. \tag{3.3}$$

Discarding terms of higher order in α_*, and taking into account (2.10), (2.15), (2.16), (2.20), (2.21), (3.1), and (3.3), we write system (II.3.4)-(II.3.14) [Eq. (II.3.5) is omitted] in the form

$$\frac{\partial \xi_r}{\partial t} + \operatorname{div} \mathbf{i}_{\xi_r} = 0, \quad r = 1, \ldots n+1, \tag{3.4}$$

$$\mathbf{i}_{\xi_r} = -\mu D_{2r} \nabla \xi_r - D_{2r+1} 2\beta_r \frac{(\xi_r)^2 \xi_{n+1}}{w} \mathbf{E}_*, \quad r = 1, \ldots n, \tag{3.5}$$

$$\mathbf{i}_{\xi_{n+1}} = -\mu D_{2n+2} \nabla \xi_{n+1} - D_{2n+3} \frac{\xi_{n+1}}{w} \mathbf{E}_* - \sum_{r=1}^{n} D_{2r+1} \beta_r \frac{(\xi_r)^2 \xi_{n+1}}{w} \mathbf{E}_*, \tag{3.6}$$

$$\operatorname{div} \mathbf{j} = 0, \tag{3.7}$$

$$\mathbf{j} = D_1 w \mathbf{E}_* + D_{2n+3} \frac{\xi_{n+1}}{w} \mathbf{E}_* + \sum_{r=1}^{n} D_{2r+1} \frac{(\xi_r)^2 \xi_{n+1}}{w} \mathbf{E}_* \beta_r, \tag{3.8}$$

$$w - \frac{\xi_{n+1}}{w} - \sum_{r=1}^{n} \beta_r \frac{(\xi_r)^2 \xi_{n+1}}{w} = 0 \quad (\xi = 0), \tag{3.9}$$

$$\frac{\partial T}{\partial t} + \operatorname{div}(-\chi \nabla T) = \delta_* (\tau_* + \mathbf{j} \cdot \mathbf{E}_*). \tag{3.10}$$

We note that, in the considered approximation, the quantities ξ_r ($r = 1, \ldots, n$), ξ_{n+1} coincide with the molar concentrations, respectively, of the polyols (D_r) and boric acid (HB) to within terms of order $O(\alpha_*)$. The system of equations (3.4)-(3.10) should be supplemented by the initial and boundary conditions. Of course, at the initial instant of time ($t = 0$), we specify the distribution of polyols and boric acid:

$$\xi_r(x, t)|_{t=0} = \xi_r^{(0)}(x), \tag{3.11}$$

$$\xi_{n+1}(x, t)|_{t=0} = \xi_{n+1}^{(0)}(x) \quad (r = 1, \ldots, n), \tag{3.12}$$

where $\xi_r^{(0)}(x)$ ($r = 1, \ldots, n$), $\xi_{n+1}^{(0)}(x)$ are the initial distributions of the polyols and the boric acid. We note that these conditions are accurate independently of α_* if when $t = 0$ the concentration of the ions H^+, B^-, $B(D_r)_2^-$ ($r = 1, \ldots, n$) are equal to zero. As the boundary conditions, let us choose the condition of impermeability of the part of the boundary Γ_0 at which the electric potential is equal to zero for all components of the solution:

$$\Gamma_0 : \mathbf{i}_{\xi_r} \cdot \mathbf{n}_0 = 0 \quad (r = 1, \ldots, n+1), \tag{3.13}$$

where $\mathbf{n}_0$ is the outward normal to Γ_0.

For the part of the boundary Γ on which the electric potential is specified, we will specify the condition of impermeability of the boundary for the neutral components of the solution:

$$\Gamma_0 : i_{2r} \cdot n = 0 \quad (r = 1, \ldots, n+1), \tag{3.14}$$

where n is the outward normal to Γ.

Neglecting terms of higher order in α_*, we obtain from (3.14), using (I.14.5), (II.3.1), (2.10), (2.15), (2.16), (2.20), (2.21), (3.1), (3.3),

$$\Gamma : \nabla \xi_r \cdot u = 0 \quad (r = 1, \ldots, n), \quad (D_{2r} \neq 0), \tag{3.15}$$

$$\Gamma : \nabla \xi_{n+1} \cdot n = 0 \quad (D_{2n+2} \neq 0). \tag{3.16}$$

From system (3.4)-(3.10), it is evident that Eq. (3.10), in the case when D_k ($k = 1, \ldots, 2n + 3$) do not depend on temperature, may be integrated if problem (3.4)-(3.9) is solved with conditions (3.11)-(3.13), (3.15), and (3.16). Therefore, in the following development we will not consider Eq. (3.10).

4. One-Dimensional Problem

In practice, a pH gradient in borate–polyol systems is created in cylindrical electrophoretic columns whose length is much greater than their diameter [50]. Therefore, it is reasonable to assume that the solution to problem (3.4)-(3.9), (3.13), (3.15), and (3.16) does not change in the radial direction, and thus to consider the one-dimensional model. Then, for convenience, we omit the subscript "*" for the quantity E and we set

$$\xi_{n+1} = b, \quad D_1 = 1, \quad D_{n+1} = D_b, \quad \xi_{n+1}^{(0)} = b^{(0)}. \tag{4.1}$$

Obviously, one of the diffusion coefficients (such as D_s) may always be set equal to unity, choosing $D_* = D_{s,\mathrm{dim}}$ as the characteristic value of the diffusion coefficient [see (II.3.1)]. It is convenient for us to set $D_* = D_{1,\mathrm{dim}}$, since the diffusion coefficient of the H^+ ion in aqueous solutions as a rule is higher than all the rest of the diffusion coefficients. In this case, for dimensionless D_s values, inequality $D_s < 1$ ($s = 2, \ldots, 2n + 3$) will be satisfied.

In the one-dimensional case, the system of equations describing the borate–polyol mixture has the form

$$b_t - \left\{ \mu D_b b_x + D_{2n+3} \frac{b}{w} E + \sum_{r=1}^{n} D_{2r+1} \beta_r \frac{(\xi_r)^2 b}{w} E \right\}_x = 0, \tag{4.2}$$

$$\xi_{r,t} - \left\{ \mu D_{2r}\xi_{r,x} + 2D_{2r+1}\beta_r \frac{(\xi_r)^2 b}{w} E \right\}_x = 0 \quad (r = 1, \ldots, n), \quad (4.3)$$

$$wE + D_{2n+3}\frac{b}{w} E + \sum_{r=1}^{n} D_{2r+1}\beta_r \frac{(\xi_r)^2 b}{w} E = j(t), \quad (4.4)$$

$$w - \frac{b}{w} - \sum_{r=1}^{n} \beta_r \frac{(\xi_r)^2 b}{w} = 0, \quad 0 \leqslant x \leqslant 1, \quad \frac{\partial}{\partial t} \equiv (\ )_t, \quad (4.5)$$

$$\frac{\partial}{\partial x} \equiv (\ )_x$$

with initial conditions

$$\xi_r(x, 0) = \xi_r^{(0)}(x), \quad r = 1, \ldots, n; \quad b(x, 0) = b^{(0)}(x) \quad (4.6)$$

and boundary conditions

$$\xi_{r,x}(0, t) = \xi_{r,x}(1, t) = 0, \quad r = 1, \ldots, n; \quad b_x(0, t) = b_x(1, t) = 0. \quad (4.7)$$

Problem (4.2)-(4.7), like (3.4)-(3.16), may be simplified. To do this, we define w and E from (4.4) and (4.5):

$$w = \sqrt{b\left(1 + \sum_{r=1}^{n} \beta_r \xi_r^2\right)}, \quad (4.8)$$

$$E = \frac{j(t)}{w + \frac{b}{w}\left(D_{2n+3} + \sum_{r=1}^{n} D_{2r+1}\beta_r \xi_r^2\right)}. \quad (4.9)$$

Eliminating from (4.2) and (4.3) the quantity

$$\frac{bE}{w} = \frac{j(t)}{(1 + D_{2n+3}) + \sum_{r=1}^{n} (1 + D_{2r+1})\beta_r \xi_r^2}, \quad (4.10)$$

we obtain the system of equations for determining the quantities $b(x, t)$ and $\xi(x, t)$:

$$b_t - \left\{ \mu D_b b_x + \frac{D_{2n+3} + \sum_{r=1}^{n} D_{2r+1}\beta_r \xi_r^2}{(1 + D_{2n+3}) + \sum_{r=1}^{n} (1 + D_{2r+1})\beta_r \xi_r^2} \cdot j(t) \right\}_x = 0, \quad (4.11)$$

$$\xi_{r,t} - \left\{ \mu D_{2r}\xi_{r,x} + \frac{2D_{2r+1}\beta_r\xi_r^2 j\,(t)}{(1+D_{2n+3}) + \sum_{r=1}^{n} (1+D_{2r+1})\,\beta_r\xi_r^2} \right\}_x = 0 \quad (r = 1,\ \dots\ n)$$

(4.12)

with initial conditions (4.6) and boundary conditions (4.7).

Usually the magnitude of the current $j(t)$ is specified. In this case, problem (4.11), (4.12), (4.6), (4.7) is closed and allows us to determine the values of the functions $b(x, t)$ and $\xi_r(x, t)$ $(r = 1, \dots, n)$. If the potential difference $U(t)$ is specified at the ends of the electrophoretic column, then system (4.11), (4.12) must be supplemented by the relation which is obtained using (4.8), (4.9):

$$U\,(t) = -\int_{0}^{1} t\,(x,\ t)\,dx,$$

(4.13)

$$j\,(t) = - \frac{U\,(t)}{\displaystyle\int_{0}^{1} \frac{\sqrt{1 + \sum_{r=1}^{n} \beta_r\xi_r^2\,(x,\ t)}}{\sqrt{b\,(x,\ t)}\left\{(1+D_{2n+3}) + \sum_{r=1}^{n}(1+D_{2r+1})\,\beta_r\xi_r^2\,(x,\ t)\right\}}\,dx}.$$

(4.14)

After determining the quantities $b(x, t)$, $\xi_r(x, t)$ $(r = 1, \dots, n)$, the functions $w(x, t)$ and $E(x, t)$ are calculated using formulas (4.8), (4.9). The pH of the solution is calculated using the formula

$$\mathrm{pH}\,(x,\ t) = 3 - \lg(\alpha_* K_*) - \lg w\,(x,\ t), \quad \dim K_* = \frac{\mathrm{moles}}{\mathrm{m}^3}.$$

(4.15)

We note that in the case where the magnitude of $j(t)$ is specified (the simpler case from the mathematical standpoint), system (4.12) may be integrated independently of Eq. (4.11). After this, the problem of determining $b(x, t)$ is reduced to solving the diffusion equation with sources.

5. Evolution of a Continuous Initial pH Gradient in the Boric Acid–Polyol System with Vanishing Diffusion

In this section, we study the possibility of creating a stable pH gradient in a system consisting of boric acid and one polyol. We consider the case where the external current in the circuit does not change direction, i.e., the direction of migration of the boric acid and polyol ions does not change. A continuous initial pH distribution along the electrophoretic column is created artificially, as described in the introductory part of this chapter.

The stability of such a pH gradient, from the standpoint of convective mixing, is ensured by the method for constructing the gradient. It is well known that the density of polyols increases as their concentration increases. Therefore, by specifying the initial pH distribution using a varying polyol concentration, we thus form some density gradient in the mixture. Setting up the electrophoretic column in such a way that the gravitational field direction coincides with the direction of increase in the density of the mixture, we create additional obstacles to the occurrence of thermal convection. Of course, the migration of ions of boric acid and polyols under the influence of an electric field leads to a redistribution of the density of the mixture. Therefore, the electrodes should be connected to the electrophoretic column in such a way that, at least in the initial stages of the process, the ion migration leads to an increase in the density gradient for the mixture; i.e., the direction of migration should coincide with the direction of increase in the polyol concentration. Rather unexpectedly, it has turned out that violation of these conditions in certain cases may lead to the appearance of discontinuities in the polyol concentration (see below). In the following development, we study the evolution of the pH gradient under the influence of an electric field under conditions when we may neglect diffusion.

Let us make the following assumptions:

$$\mu = 0, \quad n = 1, \quad j(t) \neq 0, \tag{5.1}$$

which correspond to the case of the absence of diffusion in a system consisting of boric acid and one polyol, when the direction of the external field in the circuit does not change during the isoelectric focusing process. Setting $j(t) < 0$ for concreteness, let us make the change of variables

$$\tau = \tau(t) = - \int_0^t j(s)\, ds > 0, \quad j(s) < 0,$$

$$(\ )_t = - j(t) \cdot (\ )_\tau. \tag{5.2}$$

Let us introduce the symbols

$$\alpha = \frac{2D_3}{\sqrt{(1+D_3)(1+D_5)}} > 0, \quad \alpha_1 = \frac{D_3}{1+D_3}, \quad \alpha_0 = \frac{D_5}{1+D_5},$$

$$\alpha_2^2 = \frac{1+D_5}{1+D_3}, \quad \xi_1 = \alpha_2 u, \quad p(x) = \frac{1}{\alpha_2}\xi^{(0)}(x). \tag{5.3}$$

Then system (4.11), (4.12) is rewritten in the form

$$b_\tau + \left(\frac{\alpha_0 + \alpha_1 u^2}{1 + u^2} \right)_x = 0, \tag{5.4}$$

$$u_\tau + \alpha \left(\frac{u^2}{1 + u^2} \right)_x = 0. \tag{5.5}$$

The initial conditions (4.6), taking into account (5.3), take on the form

$$b(x, t)|_{t=0} = b^{(0)}(x), \quad u(x, t)|_{t=0} = p(x). \tag{5.6}$$

The system (5.4), (5.5) is obtained from (4.11), (4.12) for the condition $\mu = 0$, i.e., it is degenerate relative to the original system of equations. Its solution may not satisfy all the boundary conditions of type (4.7). As a result, in the solution of the original problem additions of the boundary layer type may appear, compensating the discrepancies arising at the boundary. Obviously, a boundary layer for the concentration $u(x, \tau)$ will arise when $x = 1$, since the flux of the concentration $u(x, \tau)$ is oriented precisely in this direction. Therefore, as the boundary condition for (5.5) we choose the following:

$$u_x(x, \tau)|_{x=0} = 0. \tag{5.7}$$

In the following, we will not establish boundary-layer solutions, considering that the region close to the boundary is not of great practical interest.

Multiplying Eq. (5.5) by $(\alpha_0 - \alpha_1)/\alpha$ and summing with (5.4), and taking into account (5.6), we derive

$$b(x, \tau) = b^{(0)}(x) + \beta [p(x) - u(x, \tau)], \quad \beta \equiv \frac{\alpha_0 - \alpha_1}{\alpha}. \tag{5.8}$$

Thus, the quantity $b(x, \tau)$ is determined from the initial conditions if $u(x, \tau)$ is known. In order to determine $u(x, \tau)$, we have Eq. (5.5) with conditions (5.6), (5.7). Solving this by the method of characteristics (see [40, 45]), we obtain

$$u(x, \tau) = p(a(x, \tau)), \quad x > \Phi(0) \cdot \tau, \tag{5.9}$$

$$u(x, \tau) = p(a(x, 0))|_{x=0} = p(x)|_{x=0} = p(0), \quad x < \Phi(0) \cdot \tau, \tag{5.10}$$

$$\Phi(a) = \frac{2\alpha p(a)}{[1 + p^2(a)]^2}. \tag{5.11}$$

The quantity $a(x, \tau)$ is determined from the equation

$$x = \Phi(a) \cdot \tau + a, \quad a(x, 0) = x. \tag{5.12}$$

It is easy to show that in some cases, when $\tau = \tau_* < x/\Phi(0)$, formula (5.9) may give a solution which is not single-valued. The value of τ_* is determined from the condition

$$u_x(x, \tau_*) = \infty. \tag{5.13}$$

Using (5.9), (5.11), (5.12), we derive the system of equations

$$\tau_* = \frac{[1 + p^2(a)]^3}{2\alpha\,[3p^2(a) - 1] \cdot p'(a)}, \quad x = \Phi(a) \cdot \tau_* + a. \tag{5.14}$$

If system (5.14) allows for a solution satisfying the conditions

$$0 \leqslant x \leqslant 1, \quad 0 < \tau_* < \frac{x}{\Phi(0)}, \tag{5.15}$$

then, in solving problem (5.5)-(5.7), we need to introduce discontinuities. In the following development we will assume that, for the considered case, (5.13) is not satisfied; i.e., system (5.14) does not have solutions satisfying conditions (5.15). (The question concerning the discontinuous solutions will be considered in Section 7.) We note that for this it is *sufficient* (but not at all necessary) to require that one of the relations ($\tau_* < 0$) [see (5.14)] be satisfied:

$$p > \frac{1}{\sqrt{3}}, \quad p' < 0 \quad (\alpha > 0), \tag{5.16}$$

$$0 < p < \frac{1}{\sqrt{3}}, \quad p' > 0 \quad (\alpha > 0). \tag{5.17}$$

Now, for the specified functions $j(t) < 0$, using (5.2), (5.3), (5.8)-(5.11), it is easy to calculate from (4.8), (4.9), (4.13)-(4.15) the profile pH(x, t) and to determine $U(t)$ for any initial distribution $b^{(0)}(x)$ and $p(x)$, satisfying (5.16) and (5.17). For convenience, let us give the functions $w(x, \tau)$, $E(x, \tau)$ obtained from (4.8), (4.9) after the substitution (5.2), (5.3), taking into account (5.8):

$$w(x, \tau) = \sqrt{\{b^{(0)}(x) + \beta\,[p(x) - u(x, \tau)]\}\,\{1 + \alpha_2^2 u^2(x, \tau)\}}, \tag{5.18}$$

$$E(x, \tau) = \frac{j(t)}{\Lambda(u, b)},$$

$$\Lambda^{-1}(u, b) \equiv \sqrt{\frac{1 + \alpha_2^2 u^2}{b^{(0)}(x) + \beta\,[p(x) - u]}\,\frac{1}{(1 + u^2)(1 + D_5)}}, \tag{5.19}$$

where $\Lambda(u, b)$ is the conductivity of the mixture.

The solution obtained by applying formulas (4.15), (5.2), (5.3), (5.8)-(5.11), (5.18), and (5.19) has a rather unwieldy form; therefore we will restrict ourselves to establishing the asymptotic solution when $\tau \to 0$.

6. Principal Terms of the Asymptotic Expansion when $\tau \to 0$

In this section, we establish the asymptotic expansion of the solution to problem (5.4)-(5.7) when $\tau \to 0$. As we will convince ourselves in the following development, the asymptotic formulas may be successfully used for calculations in the rather broad time interval which is of practical interest.

For simplicity, let us restrict ourselves to establishing the principal terms of the asymptotic for the quantities $u(x, \tau)$, $pH(x, \tau)$. The asymptotic expansions for the quantities $w(x, \tau)$, $E(x, \tau)$, $b(x, \tau)$ are constructed analogously. We will look for $a(x, \tau)$ in the form of a series

$$a(x, \tau) = a_0(x) + a_1(x) \cdot \tau + \cdots, \quad \tau \to 0. \tag{6.1}$$

Substituting (6.1) into (5.12) and equating terms in the same powers of τ, we obtain

$$a_0(x) \equiv x, \tag{6.2}$$

$$a_1(x) = -\Phi(a_0(x)) = -\frac{2\alpha p(x)}{[1 + p^2(x)]^2}. \tag{6.3}$$

Using formulas (5.9), (5.10), we derive

$$\left.\begin{array}{l} u(x, \tau) = u_0(x) + u_1(x) \cdot \tau + \cdots, \quad \tau \cdot \Phi(0) < x, \\ u_0(x) = p(a_0(x)) = p(x), \\ u_1(x) = p'(a_0(x)), a_1(x) = -\dfrac{2\alpha p(x) p_x(x)}{[1 + p^2(x)]^2}, \\ u(x, \tau) = p(0), \quad \tau \cdot \Phi(0) > x. \end{array}\right\} \tag{6.4}$$

Finally, from (4.15) we obtain, using (6.1)-(6.4), (5.18),

$$pH(x, \tau) = pH_0(x) + pH_1(x) \cdot \tau + \ldots, \tau \to 0, \quad \tau \cdot \Phi(0) < x;$$
$$pH(x, \tau) = pH(0, 0), \quad \tau \cdot \Phi(0) > x, \tag{6.5}$$

$$pH_1(x) = \frac{\alpha p(x) p_x(x)}{b^{(0)}(x) [1 + \alpha_2^2 p^2(x)] [1 + p^2(x)]^2 \ln 10} \{2\alpha_2^2 b^{(0)}(x) p(x)$$

$$- \beta [1 + \alpha_2^2 p^2(x)]\}, \tag{6.6}$$

TABLE 19. Results of the Calculation of the Evolution of the pH Profile
for the Boric Acid–Glycerine System

x	pH value					ξ_0, moles/ liter at $t=0$
	$t=0$	$t=1$ h	$t=2$ h	$t=4$ h	$t=10$ h	
0	5.0761	5.0761	5.0761	5.0761	5.0761	0.100
0.1	4.9431	4.9811	5.0211	5.1086	5.0761	0.237
0.2	4.8100	4.8284	4.8474	4.8874	5.0761	0.376
0.3	4.6770	4.6858	4.6947	4.7131	4.7728	0.547
0.4	4.5440	4.5482	4.5525	4.5612	4.5884	0.768
0.5	4.4109	4.4130	4.4151	4.4193	4.4324	1.061
0.6	4.2778	4.2780	4.2799	4.2821	4.2885	1.454
0.7	4.1448	4.1453	4.1459	4.1469	4.1502	1.985
0.8	4.0117	4.0120	4.0123	4.0129	4.0145	2.704
0.9	3.8787	3.8788	3.8790	3.8793	3.8801	3.678
1.0	3.7456	3.7457	3.7458	3.7460	3.7464	5.000

$$pH_0(x) = 3 - \lg(\alpha_* K_*) - \frac{1}{2}\lg\{b^{(0)}(x)[1 + \alpha_2^2 p^2(x)]\}, \qquad (6.7)$$

where $pH_0(x)$ is the initial pH distribution along the electrophoretic column.

As an example, let us consider the case where the initial distribution
$pH_0(x)$ has the form of a linear pH gradient:

$$pH_0(x) = -Ax + B, \quad A > 0. \qquad (6.8)$$

Let the initial concentration of boric acid be constant:

$$b^{(0)}(x) = b^{(0)} = \text{const.} \qquad (6.9)$$

From (6.6)-(6.8), it is easy to obtain the initial distribution of the polyol
along the column which is needed to create the linear pH gradient:

$$p(x) = \frac{1}{\alpha_2}\sqrt{\frac{1}{b^{(0)}(\alpha_* K_*)^2} \cdot 10^{6-2B+2Ax} - 1} \equiv \frac{1}{\alpha_2}\xi_1^{(0)}(x). \qquad (6.10)$$

Let us give the results of the calculation of the evolution of the pH
profile for the boric acid–glycerine system. Let us choose the following
values for the parameters (in parentheses we indicate dimensionless
quantities) [7, 12]:

$$D_3 = 8.33 \cdot 10^{-10} \text{ m}^2/\text{s} \ (8.94 \cdot 10^{-2}), \quad D_* = 9.31 \cdot 10^{-9} \text{ m}^2/\text{s}, \quad L = 0.2 \text{ m (1)},$$
$$T = 552148.2 \text{ s}, \quad j(t) = -21.7 \text{ mA/cm}^2 \ (29.5223), \quad A = 1.3305, \quad B = 5.0761,$$
$$\mu = 1.29 \cdot 10^{-1}, \quad pK_1 = 7.89, \quad pK_2 = 9.24, \quad K_* = 211.349 \text{ moles/m}^3,$$
$$b^{(0)} = 0.1 \text{ moles/liter} \ (4.73 \cdot 10^{-2}), \quad 3 - \lg(\alpha_* K_*) = 4.9575,$$
$$D_5 = 2.57 \cdot 10^{-11} \text{ m}^2/\text{s} \ (2.76 \cdot 10^{-3}).$$

In Table 19 we give the results of the calculations for the instants of time $t = 1$, 2, 4, and 10 h. The pH(x, τ) values are given for points satisfying the inequality

$$\tau \cdot \Phi(0) < x. \tag{6.11}$$

The high number of significant figures in the table was necessary in order to emphasize the fact that the pH does change over the course of time. The authors admit that it is impossible to carry out the experiments with the accuracy indicated in Table 19.

The polyol concentrations for which the calculation was carried out do not satisfy requirement (5.17), which is a rather burdensome requirement in practice. As has already been indicated, this requirement is sufficient but not at all necessary. During the calculation, we directly monitored how well the inequality $\tau < \tau_*$ was satisfied [see (5.14)]; i.e., we checked the condition that the concentration profile remains continuous. For the chosen values of the concentration, the time t_* at which the first discontinuities appear is $t_* = 11$ h.

In conclusion, we note that the calculations were done for the case when the electric current density through the electrophoretic column is specified. The point is that, by specifying the potential difference at the ends of the electrophoretic column, we must inevitably take into account the falloff in potential in the regions close to the electrodes. This quantity is not very amenable to monitoring by virtue of the fact that the electrode reactions in borate–polyol systems have been studied very little. Thus, the potential difference at the ends of the column, considerably different from the potential difference within the region where the pH gradient is created, hardly characterizes the electrophoresis process. Furthermore, in specifying the electric current, in practice we are not bound to the requirements for choosing special electrodes and taking the region near the electrodes into account.

7. Evolution of a Piecewise-Constant pH Profile in the Boric Acid–Polyol System for Vanishing Diffusion

In practice, it is rather difficult to specify the necessary initial concentration of boric acid and polyol as continuous. It is much more convenient to successively transfer to the electrophoretic column some fixed volumes of the mixture with constant concentrations of boric acid and polyol (consequently, with constant pH value). Thus, at the initial instant of time, a piecewise-constant pH profile may be formed in the electrophoretic column.

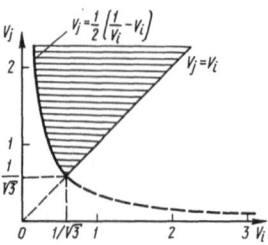

Fig. 39. Region where the concentrations must be located for the stability conditions at the discontinuities to be satisfied.

In this section we establish the generalized piecewise-constant solution to system (5.4), (5.5). We consider the case where the initial piecewise constant distribution remains piecewise-constant during the evolution process, i.e., when rarefaction waves described by variable progressive solutions do not arise.

Let us consider an infinite electrophoretic column. Assuming, as before, that there is no diffusion, we rewrite the system of equations (5.4), (5.5), describing the behavior of the boric acid–polyol mixture in the column, as follows:

$$b_\tau + \left(\frac{\alpha_0 + \alpha_1 u^2}{1 + u^2}\right)_x = 0, \tag{7.1}$$

$$u_\tau + \alpha\left(\frac{u^2}{1 + u^2}\right)_x = 0, \quad \alpha > 0. \tag{7.2}$$

Let us first consider the case where, at the instant of time $\tau = 0$, the concentration distribution has the form

$$u(x, 0) = \begin{cases} v_1, & x < a_1, \\ v_2, & a_1 < x < a_2, \\ v_3, & a_2 < x, \end{cases} \tag{7.3}$$

$$b(x, 0) = \begin{cases} c_1, & x < a_1, \\ c_2, & a_1 < x < a_2, \\ c_3, & a_2 < x. \end{cases} \tag{7.4}$$

Here, v_i, c_i ($i = 1, 2, 3$) are specified constants. Of course, then we may restrict ourselves to solution of the problem (7.2), (7.3) by determining the

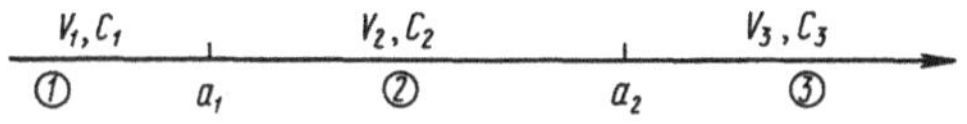

Fig. 40. Initial piecewise-constant concentration distribution of boric acid and polyol.

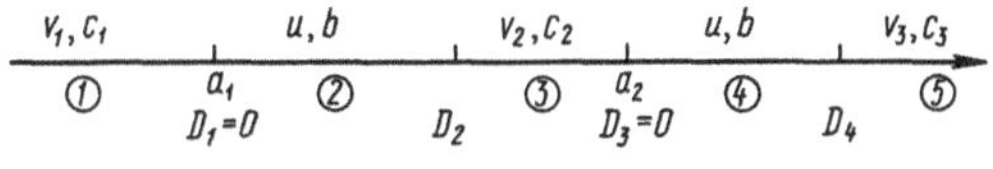

Fig. 41. Concentration distribution of boric acid and polyol, developed from the initial state and valid until the time $\tau = \tau_{23}$.

TABLE 20. Concentrations of Boric Acid and Polyol and Jump Velocities Corresponding to Fig. 41

	Zone number (k)				
	1	2	3	4	5
Polyol concentration (u)	v_1	v_1	v_2	v_2	v_3
Boric acid concentration (b)	c_1	$c_2 + \beta (v_2 - v_1)$	c_2	$c_3 + \beta (v_3 - v_2)$	c_3
Jump velocity for leading edge of k-th zone (D_k)	0	$\alpha \dfrac{v_1 + v_2}{p_1 p_2}$	0	$\alpha \dfrac{v_2 + v_3}{p_2 p_3}$	—

quantity $b(x, \tau)$ using (5.8). However, it is more expedient to make use of the results of Chapter IV, directly considering the system (7.1), (7.2) with initial conditions (7.3), (7.4).

Let us introduce the Riemann invariants, which for system (7.1), (7.2) are trivial and have the form

$$R_1 \equiv b + \beta u, \quad R_2 \equiv u, \quad \beta \equiv \frac{\alpha_0 - \alpha_1}{\alpha}. \tag{7.5}$$

The Hugoniot conditions at the discontinuities are written in the form (see Chapter IV)

$$D[b] = \left[\frac{\alpha_0 + \alpha_1 u^2}{1 + u^2} \right], \quad D[u] = \alpha \left[\frac{u^2}{1 + u^2} \right], \tag{7.6}$$

where D is the velocity of the discontinuity. Multiplying the second condition by β and summing with the first, we obtain (compare with Chapter IV)

$$D[R_1] = 0. \tag{7.7}$$

Let u_l, u_r, b_l, and b_r be the values of the concentrations to the left and to the right of the discontinuity, respectively. For fixed discontinuities, from (7.7) we derive

$$D = 0 : u_r = u_l. \tag{7.8}$$

For moving discontinuities, we obtain from (7.5), (7.6)

$$D \neq 0 : u_r \neq u_l, \quad D = \alpha \frac{u_r + u_l}{p_r p_l}, \tag{7.9}$$

$$p_r = 1 + u_r^2, \quad p_l = 1 + u_l^2, \quad b_l = b_r + \beta (u_r - u_l).$$

In Riemann invariants, system (7.1), (7.2) is rewritten in the form (compare with Chapter IV)

$$\frac{\partial R_i}{\partial \tau} = 0, \quad \frac{\partial u}{\partial \tau} + \xi \frac{\partial u}{\partial x} = 0, \quad \xi \equiv \frac{2\alpha u}{(1 + u^2)^2}. \tag{7.10}$$

The stability condition for the discontinuity has the form

$$\xi_l > D > \xi_l, \quad \frac{2\alpha u_l}{(1 + u_l^2)^3} > D > \frac{2\alpha u_r}{(1 + u_r^2)^3}. \tag{7.11}$$

Let us require that the following relations be satisfied [compare with (5.16)]:

$$v_j > v_i, \quad v_j > \frac{1}{2}\left(\frac{1}{v_i} - v_i\right), \quad j > i; \quad i, j = 1, 2, 3. \tag{7.12}$$

These conditions are sufficient in order to have the stability conditions at the discontinuities satisfied for all subsequent evolutions of the concentration profile. In Fig. 39 we show the region where the concentrations v_i, v_{i+1} must be located in order for relations (7.12) to be satisfied. We direct the attention of the reader to the fact that, in order for the stability conditions to be satisfied independently of the direction of the gravitational field, the piecewise-constant polyol concentration should increase "downstream."

Let us go directly to establishing the different stages of the evolution of the initial piecewise-constant distribution specified by (7.3), (7.4). When $\tau = 0$, the concentrations are located as follows (Fig. 40; for the meaning of the symbols see Chapter IV). At the subsequent instants of time, the zone rearrangement shown in Fig. 41 and Table 20 occurs.

Columns 1, 3, and 5 are filled in with the initial values; by virtue of the finiteness of the velocity of a disturbance for a hyperbolic system, the initial concentrations in zones 1, 3, and 5 do not change. The concentra-

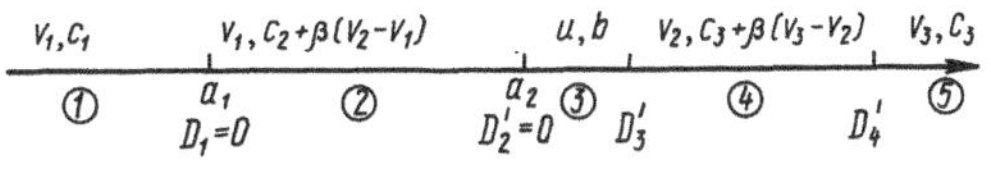

Fig. 42. Concentration distribution for boric acid and polyol in the time interval $\tau_{23} < \tau < \tau_{34}$.

TABLE 21. Concentrations of Boric Acid and Polyol and Jump Velocities Corresponding to Fig. 42

	Zone number (k)				
	1	2	3	4	5
Polyol concentration (u)	v_1	v_1	v_1	v_2	v_3
Boric acid concentration (b)	c_1	$c_2 + \beta(v_2 - v_1)$	$c_3 + \beta(v_3 - v_1)$	$c_3 + \beta(v_3 - v_2)$	c_3
Jump velocity for leading edge of kth zone (D_k)	0	0	$\alpha \dfrac{v_1 + v_2}{p_1 p_2}$	$\alpha \dfrac{v_2 + v_3}{p_2 p_3}$	—

tions u^2, u^4 in zones 2, 4 are determined from the conditions at the fixed discontinuities D_1, D_3 using (7.8):

$$u^4 = v_2, \quad u^2 = v_1. \tag{7.13}$$

From conditions (7.9) on the moving discontinuities D_2, D_4, we determine b^2, b^4 and the velocities of the discontinuities D_2, D_4:

$$\left. \begin{array}{l} b^2 = c_2 + \beta(v_2 - v_1), \quad b^4 = c_3 + \beta(v_3 - v_2), \quad D_2 = \alpha \dfrac{v_1 + v_2}{p_1 p_2}, \\[2mm] D_4 = \alpha \dfrac{v_2 + v_3}{p_2 p_3}, \quad p_i \equiv 1 + v_i^2. \end{array} \right\} \tag{7.14}$$

This pattern will be retained until the moving discontinuity D_2 crosses the fixed discontinuity D_3, which occurs at the time $\tau = \tau_{23}$:

$$\tau_{23} = \frac{a_2 - a_1}{D_2} = \frac{(a_2 - a_1)\, p_1 p_2}{\alpha(v_1 + v_2)}. \tag{7.15}$$

The zone arrangement at the time $\tau > \tau_{23}$ is shown in Fig. 42 and in Table 21. Columns 1, 2, 4, and 5 in Table 21 coincide with the corresponding columns in Table 20 by virtue of the hyperbolicity of the systems. An exception is the velocity D_2' of the leading edge of zone 2, which now, of course, is equal to zero. The concentration u^3 is determined from the condition on the fixed jump D_2':

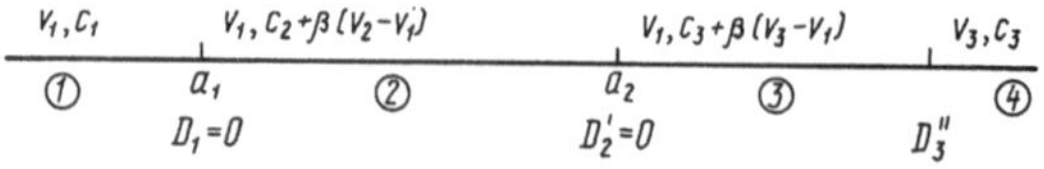

Fig. 43. Final concentration distribution for boric acid and polyol for $\tau > \tau_{34}$.

TABLE 22. Concentrations of Boric Acid and Polyol and Jump Velocities Corresponding to Fig. 43

	Zone number (k)			
	1	2	3	4
Polyol concentration (u)	v_1'	v_1	v_1'	v_3
Boric acid concentration (b)	c_1	$c_2 + \beta\,(v_2 - v_1)$	$c_3 + \beta\,(v_3 - v_1)$	c_3
Jump velocity for leading edge of kth zone (D_k)	0	0	$\alpha\,\dfrac{v_1 + v_3}{p_1 p_3}$	—

$$u^3 = v_1. \tag{7.16}$$

From the conditions on the fixed jump D_3', we find

$$b^3 = c_3 + \beta\,(v_3 - v_2) + \beta\,(v_2 - v_1) = c_3 + \beta\,(v_3 - v_1),$$
$$D_3' = \alpha\,\frac{v_1 + v_2}{p_1 p_2}. \tag{7.17}$$

Of course, on going through the fixed discontinuity, the velocity of the former zone 2 does not change: $D_2 = D_3'$. This is explained by the fact that only the concentration b "knows" about the fixed discontinuity: if we will consider Eq. (7.2), then for this equation all the discontinuities are moving.

By virtue of conditions (7.12), one more zone rearrangement occurs at the time $\tau = \tau_{34}$, when the discontinuity D_3' reaches the discontinuity D_4':

$$\tau_{34} = \frac{D_4 \tau_{23}}{D_3' - D_4}. \tag{7.18}$$

The final zone distribution pattern for $\tau > \tau_{34}$ is shown in Fig. 43 and Table 22. It is easy to show that for $\tau = \tau_{34}$, zone 4 in Fig. 42 disappears; the conditions on the discontinuities $D_2' = 0$ and $D_3'' \neq 0$ are satisfied. The concentration in columns 1-3 of Table 22 coincides with the concentrations in the corresponding tables in Table 21. Of course, in this case a change in the velocity of the discontinuity between the new zones 3 and 4 occurs. It is determined from the conditions on the discontinuity D_3'':

$$D_3'' = \alpha \frac{v_1 + v_3}{p_1 p_3}. \tag{7.19}$$

Thus, for $\tau > \tau_{34}$, a piecewise-constant concentration profile u, b, and thus a piecewise-constant pH profile, is created in the electrophoretic column. We note that, beginning from the time $\tau = \tau_{34}$, the concentration $u(x, \tau)$ and $b(x, \tau)$ between the fixed initial discontinuities a_1, a_2 is constant and remains unchanged as long as the discontinuities do not stop moving in the electrophoretic column – actually, until the moving discontinuities begin to reach the end of the electrophoretic column.

The considered example is easily generalized to the case of n initial discontinuities. For $\tau = 0$, let the distributions $u(x, 0)$ and $b(x, 0)$ have the form

$$u(x, 0) = v_i, \quad a_{i-1} < x < a_i, \quad a_0 = -\infty, \quad a_n = \infty, \tag{7.20}$$

$$b(x, 0) = c_i, \quad a_{i-1} < x < a_i, \quad i = 1, \ldots, n. \tag{7.21}$$

For v_i ($i = 1, \ldots, n$) let conditions (7.12) be satisfied. Then, for $\tau > \tau_*$, where τ_* is the time of the last possible rearrangement of the moving zones (a finite number, determined by the concentrations v_i), the concentration distribution will have the form (see Fig. 44)

$$\begin{aligned} u(x, \tau) = v_1, & \quad -\infty < x < \eta(\tau), \quad i = 1, \ldots, n-1, \\ u(x, \tau) = v_n, & \quad \eta(\tau) < x < \infty, \end{aligned} \tag{7.22}$$

$$\left. \begin{aligned} b(x, \tau) &= c_1, \quad -\infty < x < a_1, \\ b(x, \tau) &= c_i + \beta(v_i - v_1), \quad a_{i-1} < x < a_i, \quad i = 2, \ldots, n-1, \\ b(x, \tau) &= c_n + \beta(v_n - v_1), \quad a_{n-1} < x < \eta(\tau). \end{aligned} \right\} \tag{7.23}$$

Here, $\eta(\tau)$ is the location of the discontinuity D at the instant τ, $\tau > \tau_*$, and $\eta(\tau) > a_{n-1}$. In this case, the velocity of the discontinuity D is

$$D = \alpha \frac{v_1 + v_n}{p_1 p_n}. \tag{7.24}$$

Thus, for $\tau > \tau_*$, in the column there exists a stable piecewise-constant pH profile for $-\infty < x < a_{n-1}$. The pH value is easily calculated using formula (4.15):

$$pH(x, \tau) = 3 - \lg(\alpha_* K_*) - \frac{1}{2} \lg \{ [c_k + \beta(v_k - v_1)](1 + \alpha_2^2 v_1^2) \}, \tag{7.25}$$

$$a_{k-1} < x < a_k, \quad k = 1, \ldots, n-1, \quad \tau > \tau_*.$$

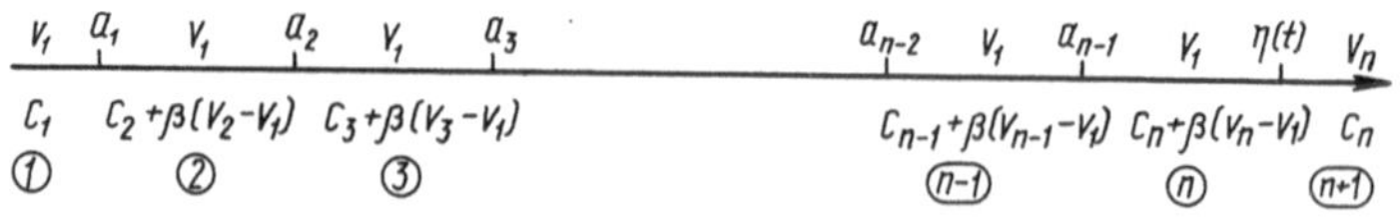

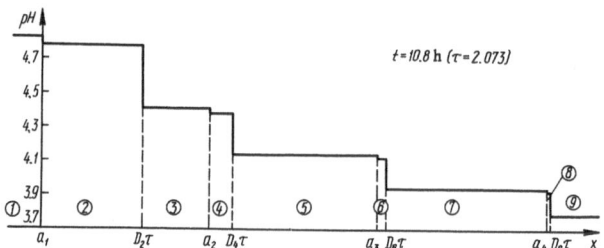

Fig. 44. Concentration distribution for boric acid and polyol in the case of n initial discontinuities after the last possible zone rearrangement.

Fig. 45. The pH distribution in the boric acid–glycerine system before the time $t = 7.43$ h.

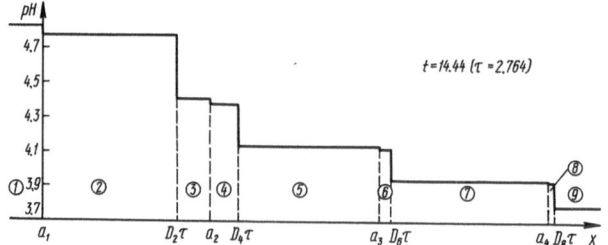

Fig. 46. The pH distribution in the boric acid–glycerine system before the time $t = 9.91$ h.

We note that the piecewise-constant pH profile is created by a piecewise-constant concentration distribution for the boric acid; the polyol concentrations for $-\infty < x < a_{n-1}$, $\tau > \tau_*$ does not change and is equal to v_1. We recall once again that such a pH distribution is retained as long as the discontinuity D is moving.

The described case of the final rearrangement of all the zones may prove to be difficult to realize in practice due to the long time τ_* needed for complete rearrangement. In fact, the maximum velocity of the discontinu-

TABLE 23. Initial Concentrations of Boric
Acid (b) and Glycerine (ξ_1), moles/liter

Investi-gated parameter	Zone number (i)				
	1	2	3	4	5
b	0.2	0.4	0.6	0,8	1.0
ξ_1	0.2	0,5	0,8	1,1	1.4

TABLE 24. Concentration Distribution for
Boric Acid (b) and Glycerine (ξ_1) Before the
Time $t = t_{min}$ = 12.35 h

i	b, moles/liter	ξ_1, moles/liter	pH	D_{i}, mm/s	$t_{2i,\,i+1}$, h
1	0.200	0.2	4.831	0	12,35
2	0,255	0.2	4.778	1,620	55,71
3	0.400	0,5	4.403	0	163.93
4	0,455	0,5	4,381	0.359	363.64
5	0,600	0,8	4.138	0	
6	0,655	0,8	4,119	0,122	
7	0.800	1,1	3,944	0	
8	0,855	1,1	3,930	0.055	
9	1,000	1,4	3.794	—	

ities split off from the initial state, for any initial distribution and satisfying
(7.12), is

$$\max_{i} \max_{(v_i v_{i+1}) \in U_i} \alpha \frac{v_{i+1} + v_i}{p_i p_{i+1}} = \frac{9\alpha}{8\sqrt{3}} \approx 0.65\alpha,$$

$$U_i = \left\{ (v_{i+1},\ v_i) : v_{i+1} > v_i,\quad v_{i+1} > \frac{1}{2}\left(\frac{1}{v_i} - v_i\right) \right\}. \tag{7.26}$$

The time needed for one such discontinuity to cross the initial discontinuity
is calculated using formula (7.15) and is

$$\tau_{**} = \frac{\Delta x \cdot 8\sqrt{3}}{9\alpha}, \tag{7.27}$$

where Δx is the maximum distance between the initial discontinuities.

In the case where (see Section 6)

$$\left.\begin{array}{l} \mathscr{T} \approx 552\ 148\ \text{s},\quad L = 0.2\ \text{m},\quad \Delta x = 0.1, \\ j = 21.7\ \text{mA/cm}^2\quad pK_1 = 7.89,\quad pK_2 = 9.24, \\ D_3 = 8.33\cdot 10^{-10}\,\text{m}^2/\text{s},\quad D_5 = 2.57\cdot 10^{-11}\,\text{m}^2/\text{s}, \end{array}\right\} \tag{7.28}$$

this time will be

$$\tau_{**} \approx 4.7 \text{ h}.$$

Obviously, the time for crossing all the zones is significantly greater than the given value. Therefore, it may turn out that, for practical purposes, the greatest interest will be in only the first zone arrangement, when the moving discontinuities starting from the fixed discontinuities have not yet crossed a single fixed discontinuity located "downstream."

The formulas for calculating the pH values in this case are given next. The initial distribution has the form

$$pH(x, \tau) = 3 - \lg(\alpha_* K_*) - \frac{1}{2}\lg[b(1 + \alpha_2^2 u^2)], \quad \tau < \tau_{**}. \quad (7.29)$$

Here

$$u(x, \tau) = \begin{cases} v_1, & -\infty < x < D_2\tau, \\ v_{i+1}, & D_{2i} \cdot \tau < x < D_{2i+2}\tau, \quad i = 1, 2, \ldots n-2, \\ v_n, & D_{2n-2} \cdot \tau < x < \infty; \end{cases} \quad (7.30)$$

$$b(x, \tau) = \begin{cases} c_1, & -\infty < x < a_1, \\ b = c_{i+1} + \beta(v_{i+1} - v_i), & a_i < x < D_{2i} \cdot \tau, \quad i = 1, \ldots n-1, \\ b = c_{i+1}, & D_{2i} \cdot \tau < x < a_{i+1}, \quad i = 1, \ldots n, \\ b = c_n, & D_{2n-2}\tau < x < \infty, \end{cases}$$

$$(7.31)$$

$$D_{2i} = \alpha \frac{v_i + v_{i+1}}{p_i p_{i+1}}, \quad i = 1, \ldots n-1, \quad (7.32)$$

$$\tau_{2i,i+1} = \frac{a_{i+1} - a_i}{D_{2i}}, \quad \tau_{**} = \min_{1 \leqslant i \leqslant n-2} \tau_{2i,i+1}. \quad (7.33)$$

Let us give an example of the calculation of a piecewise-constant pH profile using formulas (7.29)-(7.33). The parameters for the calculation are chosen to be the same as in (7.28). At the initial instant of time ($\tau = 0$), the discontinuities are distributed at 20-mm intervals at the points

$$a_1 = 0.3, \quad a_2 = 0.4; \quad a_3 = 0.5, \quad a_4 = 0.6.$$

The initial concentrations of boric acid and glycerine are given in Table 23 [see (7.20), (7.21)]. The distribution of concentrations and pH after the moving discontinuities split off from the fixed discontinuities right up to the instant of time $\tau = \tau_{**} = 12.35$ h is given in Table 24. In this table are indicated the velocities of the discontinuities and the times $\tau_{2i,i+1}$ for reaching the discontinuity D_{2i} at the point a_{i+1}, where the fixed discontinuity D_{2i+1} is located. In Figs. 45 and 46 we show the pH distribution for the times $t = 7.4$ and 9.9 h.

Chapter VII

ISOELECTRIC FOCUSING
IN INFINITE-COMPONENT MIXTURES.
CREATION OF A pH GRADIENT

This chapter is devoted to isoelectric focusing in natural pH gradients created using carrier ampholytes. The idea of natural pH gradients as an isoelectric focusing method itself is attributed to Svensson (Rilbe) (see Chapter III, and also [77, 91-93].

A pH gradient generated under the influence of an electric field (and retaining its stability over the course of a sufficiently long time interval) in an initially homogeneous mixture of substances having amphoteric properties is called a natural pH gradient. Svensson first understood the fact, which now appears obvious, that an ampholyte in the isoelectric state may have a conductivity despite the zero mobility and the fact that the overall charge is close to zero (let us recall that the values of the isoelectric and isoionic points differ slightly). Svensson also first used the concept of an ampholyte with two dissociation stages close to the isoelectric point [92].

However, realization of the idea of natural pH gradients has encountered significant difficulties related to the lack of the required carrier ampholytes: thus, in the pH range 4.0-7.0 which is important for proteins, the required carrier ampholytes generally have not been found among known chemical compounds. Convinced of the possibility of practical realization of the idea and having created natural pH gradients in oligopeptide mixtures obtained by partial hydrolysis of hemoglobin, Svensson, in 1961, formulated requirements for the ideal carrier ampholytes which, as became clear, needed to be synthesized.

The carrier ampholytes should have the following properties: high solubility in water; high buffer capacity, so that the buffer properties of the proteins or other compounds to be studied do not affect the pH gradient (even when the average carrier ampholyte is 1% and the protein concentration is 0.05%); sufficiently high conductivity, preferably homogeneous over the entire spectrum of the carriers – otherwise, in zones with considerably lower conductivity, a sharp drop in potential will occur (let us recall that a good buffer capacity is always connected with good conductivity). The carrier ampholytes should also react chemically as little as possible with the proteins (for this, in particular, they should not have any hydrophobic groups), and they should be "transparent" in the region of the UV spectrum used for spectrophotometric analysis of proteins. The carrier ampholytes should be low-molecular-weight compounds: first of all, they should not be focused too well, i.e., they should have a relatively high diffusion coefficient; second, if the proteins are contaminated by the carrier ampholytes, it will be easy to subsequently separate them. Finally, the most important requirement: in order to achieve good resolution for substances with close isoelectric points, there should be so many carrier ampholytes in the mixture that the isoelectric points of the nearest components differ by not more than 0.05 pH unit; the isoelectric points of the carrier ampholytes should span as broad a pH range as possible, preferably from 2.5 to 11.0 pH units since the isoelectric points for most proteins are found in the pH interval from 3.0 to 10.0 [84].

The problem formulated in this way was successfully solved by Vesterberg in 1969 [99]. Vesterberg's co-worker Svensson pointed out [92] that the decisive point in the successful solution of this problem was when Vesterberg noted the fact that identical protolytic groups in polyvalent acids and bases may have very different pK values. In particular, polyvalent amines have a broad spectrum of pK values. Carrying out the condensation of such amines with acrylic acid, Vesterberg obtained the required carrier ampholytes with different pK and pI values in the form of homologs and isomers of polyaminocarboxylic acids of the following type [74, 77]:

$$-CH_2-N-(CH_2)_n-N-CH_2-$$
$$\underset{\displaystyle NR_2}{\overset{\displaystyle |}{\underset{\displaystyle (CH_2)_n}{\overset{\displaystyle |}{}}}} \qquad \overset{\displaystyle |}{R}$$

where as R we may have H or $-(CH_2)_n-COOH$ ($n = 2$ or 3).

The molecular mass of these compounds, having the trade name Ampholines, ranges from 300 to 1000 D; the Swedish company LKB makes them in a form ready to use in different pH ranges. We should note that the isoelectric focusing method using ampholines has become widely used in biochemistry and molecular biology. Evidence for this is the fact that the LKB company issues a special annual bibliographic handbook, *Acta Ampholinae*, containing more than 4000 citations up to the present time. Other systems of carrier ampholytes also exist, based on a large number of related synthetic compounds. Thus, for example, Serva (West Germany) makes carrier ampholytes [74] with the trade name Servalites and formulas of the type

$$
\begin{array}{c}
\text{NH} \\
\parallel \\
H_2-N-C-NH-CH_2-N-CH_2-CH_2-CH_2-N-CH_2-CH_2-NH \\
\qquad\qquad\qquad\quad | \qquad\qquad\qquad\qquad | \qquad\qquad\quad | \\
\qquad\qquad\qquad\quad CH_2 \qquad\qquad\qquad\quad CH_2 \qquad\quad CH_2 \\
\qquad\qquad\qquad\quad | \qquad\qquad\qquad\qquad | \qquad\qquad\quad | \\
\qquad\qquad\qquad\quad CH_2 \qquad\qquad\qquad\quad CH_2 \qquad\quad PO_3H_2 \\
\qquad\qquad\qquad\quad | \qquad\qquad\qquad\qquad | \\
\qquad\qquad\qquad\quad COOH \qquad\qquad\qquad NH \\
\qquad\qquad\qquad\qquad\qquad\qquad\qquad\qquad | \\
\qquad\qquad\qquad\qquad\qquad\qquad\qquad\qquad CH_2 \\
\qquad\qquad\qquad\qquad\qquad\qquad\qquad\qquad | \\
\qquad\qquad\qquad\qquad\qquad\qquad\qquad\qquad CH_2 \\
\qquad\qquad\qquad\qquad\qquad\qquad\qquad\qquad | \\
\qquad\qquad\qquad\qquad\qquad\qquad\qquad\qquad CH_2 \\
\qquad\qquad\qquad\qquad\qquad\qquad\qquad\qquad | \\
\qquad\qquad\qquad\qquad\qquad\qquad\qquad\qquad SO_3H
\end{array}
$$

Servalites are also made for different pH ranges. This type of carrier ampholyte is obtained as a result of the reaction of several highly branched polyamides (with, on the average, 5-6 nitrogen atoms per molecule and 2-3 methylene groups between adjacent nitrogen atoms) and several derivatives containing guanidine groups with a mixture of propane sulfone and chloromethylphosphonic and acrylic acids. The components containing amino groups have pK values in the range 4.0-10.0, and those containing guanidine groups have even higher pK values. Thus, by varying the composition of the original material, we may obtain a mixture of amphoteric compounds with isoelectric points ranging from 2.0 to 11.0 [74]. At the present time, the first carrier ampholytes produced by Soviet companies have been synthesized [36].

As Svensson has demonstrated theoretically, by assuming that the pH gradient and the conductivity in the vicinity of the given zone to be focused are constant, the concentration distribution of a single carrier ampholyte is Gaussian. Using special experiments, we may investigate the actual distribution of carrier ampholytes in the pH gradient. Thus, an artificial system has been studied [75, 77] which consists of a sequence of ampholytes with identical Gaussian concentration distributions, and it has turned out that, in order to obtain a resolution of 0.02 pH unit, the system should contain not less than 20 different carrier ampholytes per pH unit.

Real mixtures of the ampholine type behave in a much more complicated manner. Investigations using ion exchange chromatography of the eluates of thin gel sections (for isoelectric focusing in polyacrylamide gel) have shown [64] that the idea of narrow zones for carrier ampholyte components is not valid in isoelectric focusing: in a 65-mm length of gel for 62 ampholine components in a 1-mm thick section, 12-14 components were observed. It is possible that in isoelectric focusing further fractionation and formation of complexes between ampholine components occurs. Many other facts are evidence for the existence of complicated and frequently unstudied electrochemical processes in which the carrier ampholytes and the substances to be separated participate [65-68]. Thus, it has been shown [91] that each carrier ampholyte arrives at the steady state with its own individual velocity; i.e., there is no synchronization. Then, in the process of "equilibrium" isoelectric focusing, a continuous shift occurs in the pH gradient ("cathode drift") until the entire solution becomes acid. Many hypotheses have been suggested to explain such instability of the pH gradient over time [87-90]; this instability, in particular, allows us to conclude that electrophoresis in this case differs from ideal isoelectric focusing, and possibly either a process of the isotachophoresis type occurs, or noncovalent interactions occur between the components (for example, association–dissociation processes).

In a number of cases, such processes also occur between the carrier ampholytes and the substances to be separated. It has been shown [78] that homogeneous macromolecules (some basic microbial proteases, beef serum albumin, myoglobin, and tRNA), forming a reversible complex with some types of carrier ampholytes in different sections of the electrophoretic chamber, give several isoelectrophoretic zones, as in the case where a preparation is used which is heterogeneous relative to the isoelectric points.

We should also note the effect of temperature on the isoelectric focusing process from the standpoint of both the increase in resolution and the optimal choice of the cooling device and thermostating accuracy. The effect of temperature on the isoelectric point is determined by the dependence $pK(T)$, which in turn depends on the carrier ampholytes, i.e., on the presence of specific protolytic groups [75]. The difference between the final temperature of isoelectric focusing and the "standard" temperature for the pH measurement should not be more than $2°$, in order to keep the variation in pI within the accuracy limit 0.01 [75].

All this shows that the isoelectric focusing process in natural pH gradients should be subjected to theoretical and experimental investigation, including mathematical modeling methods. For these purposes we may use an apparatus described in Chapters I and II as applied to infinite-component mixtures, since the number of carrier ampholytes in the described systems is high enough for us to use such an approximation.

It is most convenient to characterize mixtures with a large number of carrier ampholytes by a continuous sorting parameter. As the sorting parameter we may use the length of the polymeric chains in the mixture or the dissociation constants of the mixture. In the mathematical description of the mixture, it is possible to separate out the closed chemical subsystems whose mobility and molar charge depend only on the sorting parameter and the pH of the medium. Mixtures of carrier ampholytes should have the following properties. Isoelectric pH values should exist for which the mobility of the closed chemical subsystems goes to zero. In this case, the isoelectric values should not coincide with the isoionic values for which the charge of the closed chemical subsystem goes to zero. In this case, while migrating under the influence of an electric field, each closed chemical subsystem should stop at the point at which the pH value of the medium coincides with the isoelectric value (the existence of an initial pH gradient is assumed). In this case, some free charge distribution is created relative to the solution, since the molar charge at the isoelectric point is different from zero. For a solution which is close to electrically neutral, the charge generated is compensated by the charge of the water ions. Thus, some pH distribution may arise in the solution.

In Section 1 of this chapter the equations for describing electrophoresis processes are generalized for infinite-component mixtures (see Chapter I), whose properties depend on a continuous sorting parameter. It is assumed that there are no fluxes in the sorting parameter space. In Section 2 we formulate the problem of determining the properties of the mixture, the electric field intensity, the temperature, etc., for a given pH profile. In Sections 3 and 4 we establish the principal term in the asymptotic expansion of the solution for the small diffusion parameter ($\mu \to 0$) for solutions which are close to electrically neutral ($\theta \to 0$). Then, Sections 5-7 are devoted to constructing different models for the mixture in which it is possible to create a linear pH profile. These models differ from one another only in the choice of a sorting parameter and the nature of the functional dependence of the difference between the isoelectric and the isoionic value on the sorting parameter. A specific variant of this dependence is chosen from consideration of convenience, taking into account available experimental information. The calculations given in Section 8 allow us to conclude that these models give results which are qualitatively the same. Experiment should play the deciding role in choosing a specific model.

The temperature distribution in an infinite-component mixture is studied in Section 9. The effect of the cooling device may be modeled by different methods. In the case when we may neglect transverse heat fluxes, it is most expedient to model the cooling device by a volume heat sink. Transverse heat fluxes will significantly differ from zero when the diameter (or the width) of the electrophoretic chamber is comparable with its length. In this case, a strong change in temperature may arise in the transverse direction, and we may more reasonably model the cooling device by specifying the heat fluxes on the lateral boundaries of the chamber.

1.   Description of Infinite-Component Mixtures

In this section we construct a mathematical model for a mixture of chemical subsystems when the number of mixture components is sufficiently high. As has been indicated in Chapter I, a promising way to describe the system in this case is connected with introducing a model for a mixture consisting of an infinitely high number of components. Then we will consider a mixture composed of a solvent (water), its ions, and an infinite number of closed chemical subsystems. We will characterize the individual closed chemical subsystems by the sorting parameter k. The correct choice of the sorting parameter requires certain information about the physicochemical properties of the mixture and allows for some arbitrariness. For linear polymers it is reasonable to take the length of the polymeric chain or the distance between its ends as k (see [2, 8]). In a number of cases we need to consider a multidimensional sorting parameter – for example, the set of dissociation constants in the case of polyelectrolytes.

In order to describe an infinite-component mixture, let us introduce the concentration distribution function for the components of the mixture, $\xi(k, x, t)$, depending on the multidimensional sorting parameter k. The composition of the mixture occupying the volume R at the initial instant of time is specified by the global distribution function

$$M(k) = \int_R \xi(k, x, 0)\, dv. \qquad (1.1)$$

As the mathematical model describing the behavior of an infinite-component mixture, let us choose the model given in Section 4 of Chapter II. Let us assume that there are no fluxes in the sorting space. Then the generalization of this model to the case of an infinite-component mixture involves the natural replacement of the summation over the index k by integration over the parameter k in the region of variation of the parameter G. Of course, the described transition, as noted in Chapter I, is a postulate.

The physicochemical properties of such a mixture will be described using the distribution functions of the diffusion coefficient $D(k, \psi(x, t))$, the mobilities $\gamma(k, \psi(x, t))$, the conductivities $\sigma(k, \psi(x, t))$, the diffusion conductivity $\sigma^D(k, \psi(x, t))$, and the molar charge $e(k, \psi(x, t))$. Let us recall (see Chapter II) that, for closed chemical subsystems, the diffusion coefficient, mobility, etc., depend only on the acidity of the mixture ($\psi(x, t)$).

Then, let us restrict ourselves to the case when the mixture as a whole is at rest, the density of the mixture is constant, there are no barodiffusion and thermal diffusion effects, the dielectric constant of the mixture equals the dielectric constant of the solvent (for the case of polarizable media, see [17]), and the diffusion coefficients and the mobilities of the individual components do not depend on temperature (see Section 2 of Chapter III). Furthermore, we will assume that the mixture of chemical subsystems is electrically neutral:

$$\theta = 0. \tag{1.2}$$

Let us write the equations describing the behavior of an infinite-component mixture [compare with (II.4.13)-(II.4.19)]:

$$\frac{\partial \xi (k, x, t)}{\partial t} + \operatorname{div} i (k, x, t) = 0, \tag{1.3}$$

$$i (k, x, t) = - \mu \nabla \{ D (k, \psi) \xi (k, x, t) \} + \gamma (k, \psi) \xi (k, x, t) E (x, t), \tag{1.4}$$

$$\operatorname{sh} \psi (x, t) + \int_G e (k, \psi) \xi (k, x, t) \, dk = 0, \tag{1.5}$$

$$j (t) = \{ - \mu \nabla \psi (x, t) + E (x, t) \} \operatorname{ch} \{ \psi (x, t) - \psi_* \}$$
$$+ \int_G \{ - \mu \nabla [\gamma (k, \psi) \xi (k, x, t)] + \sigma (k, \psi) \xi (k, x, t) E (x, t) \} \, dk. \tag{1.6}$$

Let us recall that, by virtue of the assumptions made, the thermal conductivity equations may be integrated after solving system (1.3)-(1.6), and, therefore, we will not write them out here.

2. Formulation of the Problem of Creation of the pH Gradient

We will use the system of equations presented in the preceding section to construct a model for creation of the pH gradient in the mixture. The following steady-state problem is of practical interest in using the isoelectric focusing method. For a specified monotonic acidity profile $\psi(x)$ [$\partial \psi(x)/\partial x \neq 0$] for a known set of amphoteric substances characterized by the quantities $D(k, \psi)$, $\gamma(k, \psi)$, $\sigma^D(k, \psi)$, $\sigma(k, \psi)$, $e(k, \psi)$, we need to determine the composition of the mixture, i.e., the global distribution $M(k)$, the concentration distribution $\xi(\mathbf{k}, x)$, and the electric field intensity. We assume that the potential V or the electric current density $\mathbf{j}$ is specified.

Let us consider the one-dimensional steady-state case. Let the mixture of chemical subsystems be placed in an electrophoretic column with cross section S ($0 \leq x \leq 1$). Let us rewrite system (1.3)-(1.6) assuming that the parameter k is one-dimensional:

$$\frac{\partial i(k, x)}{\partial x} \equiv \frac{\partial}{\partial x} \left\{ -\mu \frac{\partial}{\partial x} [D(k, \psi) \xi(k, x)] + \gamma(k, \psi) \xi(k, x) E(x) \right\} = 0,$$

$$k \in G \qquad\qquad (2.1)$$

$$\text{sh } \psi(x) + \int_G e(k, \psi(x)) \xi(k, x) \, dk, \qquad\qquad (2.2)$$

$$j = \left\{ -\mu \frac{\partial \psi(x)}{\partial x} + E(x) \right\} \text{ch } \{\psi(x) - \psi_*\}$$

$$+ \int_G \left\{ -\mu \frac{\partial}{\partial x} [\gamma(k, \psi) \xi(k, x)] + \sigma(k, \psi) \xi(k, x) E(x) \right\} dk. \qquad (2.3)$$

Let us assume that there is no flux of the kth chemical subsystem through the boundary:

$$i(k, 0) = i(k, 1) = 0, \quad k \in G. \qquad\qquad (2.4)$$

The global distribution $M(k)$, i.e., the composition of the mixture, is determined by the expression [it is assumed that $M = O(1)$]

$$M(k) = S \int_0^1 \xi(k, x) \, dx, \quad k \in G. \qquad\qquad (2.5)$$

The potential between the ends of the electrophoretic column is

$$V = \int_0^1 E(x) \, dx. \qquad\qquad (2.6)$$

Let the monotonic acidity profile $\psi(x)$ and the current density j (or the quantity V) be specified. Then it turns out that system (2.1)-(2.3) with conditions (2.4) allows us to determine the quantities $\xi(k, x)$, $M(k)$, and $E(x)$ if the infinite-component mixture has amphoteric properties. The latter means that, for any $k \in G$, the following relation should be satisfied:

$$\gamma(k, \psi'(k)) = 0, \quad k \in G, \qquad\qquad (2.7)$$

where $\psi^i(k)$ is the distribution of isoelectric values in the mixture relative to the sorting parameter k.

3. Establishing the Principal Term in the Asymptotic Expansion for $\mu \to 0$

Let us assume that the distribution of the ampholyte mobilities monotonically depends on the acidity:

$$\gamma_\psi (k, \psi) > 0, \quad (\ )_\psi \equiv \frac{\partial (\ )}{\partial \psi}, \quad k \in G. \tag{3.1}$$

Let the monotonic function $\psi(x)$ be specified. For concreteness, we set

$$\psi_x (x) < 0, \quad x \in [0, 1], \quad (\ )_x \equiv \frac{\partial (\ )}{\partial x}. \tag{3.2}$$

Let us introduce the distribution of isoelectric points in the mixture $x = \alpha(k)$ with the requirement (compare with Section 1 of Chapter III)

$$\gamma (k, \psi (\alpha (k))) \equiv 0, \quad k \in G. \tag{3.3}$$

We note that the value $\alpha(k)$ corresponds to the point $x \in [0, 1]$ at which the mobility of the kth chemical subsystem goes to zero.

We analogously determine the inverse function $k = \beta(x)$, indicating the value of the sorting parameter for which the mobility at the point goes to zero:

$$\gamma (\beta (x), \psi (x)) \equiv 0, \quad x \in [0, 1]. \tag{3.4}$$

Let us write the following obvious relations:

$$\left. \begin{array}{l} \alpha (\beta (x)) \equiv 1, \quad x \in [0, 1], \\ \beta (\alpha (k)) \equiv 1, \quad k \in G, \\ \alpha_k (k) \beta_x (x) \equiv 1, \quad (\ )_k \equiv \frac{\partial}{\partial k} (\ ). \end{array} \right\} \tag{3.5}$$

Let us make the assumption, which can be verified *a posteriori*, that the electric field intensity in the considered volume does not change sign (for concreteness, let it be positive); this is easy to achieve by changing, if necessary, the polarity of the applied electric field:

$$E (x) > 0, \quad x \in [0, 1]. \tag{3.6}$$

Integrating (2.1) while taking into account (2.4), we obtain

$$\xi(k, x) = \xi(k, \alpha(k)) \frac{D(k, \psi(\alpha(k)))}{D(k, \psi(x))} \exp\left\{\frac{1}{\mu} \int_{\alpha(k)}^{x} \frac{\gamma(k, \psi(s))}{D(k, \psi(s))} E(s)\, ds\right\}. \quad (3.7)$$

Let us introduce the symbol

$$U(k, x) = \int_{\alpha(k)}^{x} \frac{\gamma(k, \psi(s))}{D(k, \psi(s))} E(s)\, ds, \quad k \in G, \quad x \in [0, 1]. \quad (3.8)$$

From (3.8), using (3.3), we derive

$$U_x(k, x) = \frac{\gamma(k, \psi(x))}{D(k, \psi(x))} E(x), \quad k \in G, \quad x \in [0, 1], \quad (3.9)$$

$$U_{xx}(k, x) = \frac{\gamma_\psi \psi_x E + \gamma E_x}{D} - \frac{\gamma D_\psi \psi_x E}{D^2}, \quad k \in G, \quad x \in [0, 1], \quad (3.10)$$

$$U_k(k, x) = \int_{\alpha(k)}^{x} \left(\frac{\gamma}{D}\right)_k E\, ds - \frac{\gamma(k, \psi(\alpha(k)))}{D(k, \psi(\alpha(k)))} E(\alpha(k)) \alpha_k'(k) = \int_{\alpha(k)}^{x} \left(\frac{\gamma}{D}\right)_k E\, ds, \quad (3.11)$$

$$U_{kk}(k, x) = \int_{\alpha(k)}^{x} \left(\frac{\gamma}{D}\right)_{kk} E\, ds - \left(\frac{\gamma}{D}\right)_k E\big|_{x=\alpha(k)} \cdot \alpha_k', \quad (3.12)$$

from which, using (3.1)-(3.6), (3.8), we obtain

$$\left.\begin{aligned}
&U(k, x) \leqslant 0, \quad k \in G, \quad x \in [0, 1], \quad U(k, \alpha(k)) = 0, \\
&U_x(k, \alpha(k)) = 0, \quad k \in G, \\
&U_{xx}(k, \alpha(k)) = \frac{\gamma_\psi(k, \psi(\alpha(k))) \psi_x(\alpha(k)) E(\alpha(k))}{D(k, \psi(\alpha(k)))} < 0, \quad k \in G,
\end{aligned}\right\} \quad (3.13)$$

$$\left.\begin{aligned}
&U(\beta(x), x) = 0, \quad U_k(\beta(x), x) = 0, \quad x \in [0, 1], \\
&U_{kk}(\beta(x), x) = -\frac{\gamma_k(\beta(x), \psi(x)) E(x)}{D(\beta(x), \psi(x)) \beta_x(x)} < 0, \quad x \in [0, 1].
\end{aligned}\right\} \quad (3.14)$$

The inequality $U_{kk}(\beta(x), x) < 0$ follows from the validity of $U(k, x) \leq 0$ for $k \in G, x \in [0, 1]$. Substituting (3.7) into (3.5) and using (3.13), by the Laplace method we derive (see, for example, [53])

$$M(k) = S \sqrt{\frac{2\pi\mu}{-U_{xx}(k,\,\alpha(k))}} \{\xi(k,\,\alpha(k)) + O(\sqrt{\mu})\}, \quad \mu \downarrow 0, \quad k \in G, \quad (3.15)$$

from which we get

$$\xi(k,\,\alpha(k)) = \frac{M(k)\sqrt{-U_{xx}(k,\,\alpha(k))}}{S\sqrt{2\pi\mu}} + O(\sqrt{\mu}), \quad \mu \downarrow 0, \quad k \in G. \quad (3.16)$$

Let us recall that $M(k) = O(1)$.

Let us write the principal term of the asymptotic expansion for the following expression:

$$\int_G e(k,\,\psi(x))\,\xi(k,\,x)\,dk = \int_G e(k,\,\psi(x))\,\xi(k,\,\alpha(k))\,\frac{D(k,\,\psi(\alpha(k)))}{D(k,\,\psi(x))}$$

$$\times \exp\left(\frac{1}{\mu}\,U(k,\,x)\,dk\right) = e(\beta(x),\,\psi(x))$$

$$\times \left\{\frac{M(\beta(x))\sqrt{-U_{xx}(\beta(x),\,x)}}{S\sqrt{2\pi\mu}} + O(\sqrt{\mu})\right\} \frac{\sqrt{2\pi\mu}}{\sqrt{-U_{kk}(\beta(x),\,x)}}, \quad (3.17)$$

$$\beta(x) \in G,$$
$$\beta(x) \bar\in \partial G.$$

Using (3.13), (3.14), we obtain

$$\int_G e(k,\,\psi(x))\,\xi(k,\,x)\,dk = e(\beta(x),\,\psi(x))$$

$$\times \left\{\frac{M(\beta(x))}{S}\sqrt{-\frac{\gamma_\psi(\beta(x),\,\psi(x))\,\psi_x(x)\,\beta_x(x)}{\gamma_k(\beta(x),\,\psi(x))}} + O(\mu)\right\} \quad (3.18)$$

Differentiating (3.4) with respect to x, we have

$$\gamma_k\beta_x + \gamma_\psi\psi_x \equiv 0, \quad (\gamma_\psi(\beta(x),\,\psi(x))\,\psi_x(x) = -\gamma_k(\beta(x),\,\psi(x))\,\beta_x(x)) \quad (3.19)$$

From this we have

$$\int_G e(k,\,\psi(x))\,\xi(k,\,x)\,dk = e(\beta(x),\,\psi(x))\left\{\frac{M(\beta(x))\,|\beta_x(x)|}{S} + O(\mu)\right\},$$
$$\mu \downarrow 0, \quad \beta(x) \in G, \quad \beta(x) \bar\in \partial G. \quad (3.20)$$

Analogously,

$$\int_G \sigma(k, \psi(x))\, \xi(k, x)\, dk = \sigma(\beta(x), \psi(x)) \left\{ \frac{M(\beta(x))\,|\,\beta_x(x)\,|}{S} + O(\mu) \right\}, \quad (3.21)$$

$$\int_G \left\{ -\mu \frac{\partial}{\partial x}\, [\gamma(k, \psi)\, \xi(k, x)]\, dk \right\} = -\mu \gamma_\psi (\beta(x), \psi(x))\, \psi_x(x)$$
$$\times \left\{ \frac{M(\beta(x))\,|\,\beta_x(x)\,|}{S} + O(\mu) \right\}. \quad (3.22)$$

Substituting (3.20)-(3.22) into (2.2), (2.3), we write

$$0 = \mathrm{sh}\,\psi(x) + e(\beta(x), \psi(x))\, \{M(\beta(x))\,|\,\beta_x(x)\,| + O(\mu)\}, \quad (3.23)$$

$$E(x) = \frac{j + \mu\psi_x(x)\,\mathrm{ch}\,\{\psi(x) - \psi_*\} + \mu\gamma_\psi(\beta(x), \psi(x))\,\psi_x(x)\,\{M(\beta(x))\,|\,\beta_x(x)\,| + O(\mu)\}}{\mathrm{ch}\,\{\psi(x) - \psi_*\} + \sigma(\beta(x), \psi(x))\,\{M(\beta(x))\,|\,\beta_x(x)\,| + O(\mu)\}}.$$
$$(3.24)$$

4. Principal Term of the Asymptotic for a Mixture of Carrier Ampholytes

The results obtained in the preceding section are valid for any mixture having amphoteric properties. Below we consider the case of the so-called mixture of carrier ampholytes, for example polyaminocarboxylic acids. We will describe the properties of the mixture using the distribution functions for the diffusion coefficient, the mobility, the molar electric conductivity, and the molar charge [compare with (III.4.12)]:

$$D(k, \psi) = \frac{m(k)\, D^0(k) + \sqrt{D^-(k)\, D^+(k)}\ \mathrm{ch}\,(\psi - \psi^i(k))}{\mathrm{ch}\,(\psi - \psi^e(k)) + m(k)}, \quad (4.1)$$

$$\gamma(k, \psi) = D^-(k)\, e^{\psi^e(k) - \psi^i(k)}\ \frac{\mathrm{sh}\,(\psi - \psi^i(k))}{\mathrm{ch}\,(\psi - \psi^e(k)) + m(k)}, \quad (4.2)$$

$$e(k, \psi) = \frac{\mathrm{sh}\,(\psi - \psi^e(k))}{\mathrm{ch}\,(\psi - \psi^e(k)) + m(k)}, \quad (4.3)$$

$$\sigma(k, \psi) = D^-(k)\, e^{\psi^e(k) - \psi^i(k)}\ \frac{\mathrm{ch}\,(\psi - \psi^i(k))}{\mathrm{ch}\,(\psi - \psi^e(k)) + m(k)}. \quad (4.4)$$

Here, $D^-(k)$ is the distribution of the diffusion coefficients for the negative ions with charge -1 relative to the sorting parameter, $\psi^e(k)$ is the distribution of isoionic values relative to the sorting parameter (let us recall that we have designated as the isoionic value that value of the acidity at which the charge goes to zero); $\psi^i(k)$ is the distribution of isoelectric points relative to the sorting parameter; $m(k)$ is the dependence of the ratio of the dissociation constants on the sorting parameter.

Relations (4.2)-(4.4) correspond to an amphoteric substance dissociating in aqueous solutions in two stages. If we know the dependence of the dissociation constant on the sorting parameter, then

$$m(k) = \frac{1}{2} \sqrt{\frac{K^+(k)}{K^-(k)}}, \quad \psi^e(k) = \ln 10^7 \sqrt{K^-(k) K^+(k)},$$

$$\psi^i(k) - \psi^e(k) = \frac{1}{2} \ln \frac{D^-(k)}{D^+(k)}, \quad K^-(k), \quad K^+(k) = \left[\frac{moles}{liter}\right],$$

$$(4.5)$$

where $K^+(k)$, $K^-(k)$ are the dissociation constants for the ammonium and carboxyl groups, respectively; $D^+(k)$ is the distribution of the diffusion coefficients for the positive ions with charge +1 relative to the sorting parameter.

We note that the values of the functions (4.1)-(4.4) are necessary only at the isoelectric points. Therefore, to a sufficient degree of accuracy, the relations are satisfied even for amphoteric substances dissociating in many stages, since usually the difference between the dissociation constants for the different stages is sufficiently high, and the dissociation in the vicinity of the isoelectric point occurs as in two-stage amphoteric substances. In this case, $K^+(k)$, $K^-(k)$ correspond to the dissociation constants for dissociation to monovalent positive and negative ions (also see Section 6 of Chapter III).

The functions $x = \alpha(k)$, $k = \beta(x)$ are now determined by the relations

$$\psi(\alpha(k)) = \psi^i(k), \psi(x) = \psi^i(\beta(x)). \tag{4.6}$$

Then, from (4.1)-(4.4) we derive

$$\sigma(k, \psi^i(k)) = \gamma_\psi(k, \psi^i(k)) = \sqrt{D^-(k) D^+(k)}$$

$$\times \frac{1}{ch\{\psi^i(k) - \psi^e(k)\} + m(k)} > 0, \tag{4.7}$$

$$e(k, \psi^i(k)) = \frac{sh(\psi^i(k) - \psi^e(k))}{ch(\psi^i(k) - \psi^e(k)) + m(k)}. \tag{4.8}$$

Let us assume that

$$\psi_x(x) = 0\,(1), \quad x \in [0, 1]. \tag{4.9}$$

Then, using (4.7), we rewrite (3.24) in the form

$$E(x) = \frac{j}{\operatorname{ch}(\psi - \psi_*) + \sigma\,\dfrac{M\,|\beta_x|}{S}} + \frac{\mu\psi_x}{1 + \dfrac{\sigma M\,|\beta_x|}{S\,\operatorname{ch}(\psi - \psi_*)}}$$

$$+ \frac{\mu\psi_x}{1 + \dfrac{S\,\operatorname{ch}(\psi - \psi_*)}{\sigma M\,|\beta_x|}} = \frac{j}{\operatorname{ch}\{\psi(x) - \psi_*\} + \sigma\,(\beta(x),\,\psi(x))\,\dfrac{M\,(\beta(x))\,|\beta_x(x)|}{S}}$$

$$+ O(\mu) > 0, \quad x \in [0,1]. \tag{4.10}$$

We note that sign $E = \operatorname{sign} j$, and condition (3.6) is easy to verify. Then let

$$E_x(x) = O(1), \quad x \in [0, 1], \quad \psi^i(k) \not\equiv \psi^e(k);$$
$$e(k,\,\psi^i(k)) = O(1), \quad k \in G. \tag{4.11}$$

In this case, we rewrite (3.23) in the form

$$\operatorname{sh}\psi(x) + e(\beta(x),\,\psi(x))\,\frac{M(\beta(x)) \cdot |\beta_x(x)|}{S} = O(\mu), \quad x \in [0, 1]. \tag{4.12}$$

The relations obtained, together with (4.6)-(4.8), allow us to easily calculate the dependence $M(k)$ for the specified profile

$$M(k) = -\frac{S\,|\alpha_k(k)|\,\operatorname{ch}\psi^i(k)\,\{\operatorname{ch}(\psi^i(k) - \psi^e(k)) + m(k)\}}{\operatorname{ch}\{\psi^i(k) - \psi^e(k)\}}. \tag{4.13}$$

5. Creation of a Linear pH Profile (First Model)

As an example, let us consider the creation of a linear pH gradient. Let us choose the function $\psi(x)$ in the form

$$\psi(x) = -ax + \psi_0, \quad \psi_0 \equiv \psi(0), \quad a > 0, \quad x \in [0, 1]. \tag{5.1}$$

Let us introduce the reduced dissociation constants

$$pK^+ = -\lg K^+, \quad pK^- = -\lg K^-. \tag{5.2}$$

From (4.5), (4.6) we obtain

$$\psi^e(k) = \left(7 - \frac{pK^+ + pK^-}{2}\right)\ln 10, \quad m(k) = \frac{1}{2}\,10^{\frac{pK^- - pK^+}{2}}. \tag{5.3}$$

Going to pH values, we have

$$pI = \frac{pK^+ + pK^-}{2}, \quad m(k) = \frac{1}{2} 10^{pI - pK^+}, \quad K_w = 10^{-7} \frac{moles}{liter}, \quad (5.4)$$

where pI is the isoelectric point in pH units. The quantity pI $-$ pK$^+$ is one of the determining characteristics of the carrier ampholytes (see, for example, [46]).

Differentiating (4.3) with respect to ψ, we obtain

$$e_\psi(k, \psi) = \frac{1 + m(k) \, ch(\psi - \psi^e(k))}{[ch(\psi - \psi^e(k)) + m(k)]^2}, \quad e_\psi(k, \psi^e(k)) = \frac{1}{1 + m(k)}. \quad (5.5)$$

From this it is evident that in order for the carrier ampholyte to have a narrow pH interval for the isoionic value [46], i.e., in order for the quantity $e_\psi(k, \psi^e(k))$ to be sufficiently high, we need to require that the quantity $M(k)$ be as small as possible:

$$\max_k e_\varphi(k, \psi^e(k)) = \frac{1}{1 + \min m(k)} = \frac{1}{1 + \frac{1}{2} 10^{\min(pI - pK^+)}},$$

$$pI - pK^+ > 0. \quad (5.6)$$

It has been experimentally established that the most desirable value for pI $-$ pK$^+$ is 1.5 [46].

Next, let us consider an infinite-component mixture for which the quantity pI $-$ pK$^+$ is fixed:

$$pI - pK^+ = G = const. \quad (5.7)$$

As the sorting parameter k, let us choose the quantity

$$k \equiv pI. \quad (5.8)$$

Then, from (4.5)-(4.8), we derive

$$\left. \begin{array}{l} \psi^i(k) = (7 - k) \ln 10 + \delta_0(k), \quad \psi^e(k) = (7 - k) \ln 10, \quad m(k) \\ = \frac{1}{2} 10^G = const, \quad \alpha(k) \equiv \frac{1}{a} \{\psi_0 - (7 - k) \ln 10 - \delta_0(k)\}, \\ \beta(x) \equiv 7 + \frac{ax - \psi_0 + \delta_0(\beta(x))}{\ln 10}, \quad e(k, \psi^i(k)) \equiv \frac{sh \, \delta_0(k)}{ch \, \delta(k) + m(k)}, \\ \sigma(k, \psi^i(k)) \equiv \sqrt{D^-(k) D^+(k)} \frac{1}{ch \, \delta_0(k) + m(k)}, \\ \delta_0(k) \equiv \frac{1}{2} \ln \frac{D^-(k)}{D^+(k)}, \quad D^0(k) \equiv \sqrt{D^-(k) D^+(k)}. \end{array} \right\} \quad (5.9)$$

For simplicity, let us consider the case when

$$\delta_0(k) \equiv \text{const}, \quad D^0(k) \equiv \text{const}. \tag{5.10}$$

We note that the quantities $\delta_0(k)$, $D^0(k)$ may be found from an experiment determining the migration rate of the components in an electric field. At the present time these quantities have not been determined.

If condition (5.10) is satisfied, then

$$\alpha_k(k) = \frac{\ln 10}{a}, \tag{5.11}$$

$$\beta_x(x) = \frac{a}{\ln 10}, \quad \frac{\delta_0 - \psi_0}{\ln 10} + 7 \leqslant k \leqslant \frac{a + \delta_0 - \psi_0}{\ln 10} + 7.$$

For the distribution of the amount of the substance relative to the sorting parameter and the electric field intensity, let us derive from (4.10), (4.13)

$$M(k) = - \frac{S \dfrac{\ln 10}{a} \operatorname{sh}\{(7-k)\ln 10 + \delta_0\}(\operatorname{ch}\delta_0 + m)}{\operatorname{sh}\delta_0} \geqslant 0, \tag{5.12}$$

$$E(x) = \frac{}{\operatorname{ch}(\psi(x) - \psi_*) - \dfrac{D^0}{\operatorname{sh}\delta_0}\operatorname{sh}\psi(x)} > 0, \quad j > 0. \tag{5.13}$$

We note that the condition $M(k) \geq 0$ imposes restrictions on the interval of variation in $\psi(x)$. For the chosen model, we need to require

$$\psi(x) \cdot \delta_0 < 0, \quad x \in [0, 1]. \tag{5.14}$$

This means that, in the case $\delta_0 = \text{const}$, the quantity $\psi(x)$ should not go to zero. In other words, it is possible to create a pH gradient only with pH < 7 ($\delta_0 < 0$) or pH > 7 ($\delta_0 > 0$).

Introducing the symbols

$$c = \operatorname{ch}\psi_*, \quad b = -\operatorname{sh}\psi_* - \frac{D^0}{\operatorname{ch}\delta_0}, \tag{5.15}$$

let us rewrite (5.13) in the form

$$E(x) = \frac{j}{c\operatorname{ch}\psi + b\operatorname{ch}\psi}. \tag{5.16}$$

Substituting the expression obtained into (2.6), using (5.1) we obtain

$$V = -\frac{1}{a} \int_{\psi(0)}^{\psi(1)} \frac{j\,d\psi}{c\,\text{ch}\,\psi + b\,\text{sh}\,\psi} = -\frac{j}{a\sqrt{b^2-c^2}} \ln \left| \frac{\text{th}\,\dfrac{\psi(1) + \text{Arth}\,\dfrac{c}{b}}{2}}{\text{th}\,\dfrac{\psi(0) + \text{Arth}\,\dfrac{c}{b}}{2}} \right|, \tag{5.17}$$

$$b^2 > c^2.$$

The latter expression allows us to determine the quantity j using known V.

We note that the quantity $M(k)$ should satisfy the inequality

$$M(k) \leqslant M_{\text{sol}}(k), \quad k \in G, \tag{5.18}$$

where $M_{\text{sol}}(k)$ is the solubility of components of the kth type. Condition (5.18) may be satisfied as a result of the choice of the cross section S.

6. Creation of a Linear pH Profile (Second Model)

In the preceding section the basic characteristic of the infinite-component mixture was the quantity $pI - pK^+ = G$ [see (5.7)]. Obviously, we may construct other models for the mixture, choosing other basic parameters (other sorting parameters). For example, let us consider a mixture for which the dissociation constants pK^+ of the most acid group is fixed:

$$pK^+ = \text{const.} \tag{6.1}$$

As the sorting parameter, let us choose the ionization constant pK^-:

$$k \equiv pK^-. \tag{6.2}$$

Specifying the profile $\psi(x)$ in the form (5.1), and setting as before

$$\delta_0(k) = \text{const}, \ D^0(k) = \text{const}, \ m(k) = \frac{1}{2} 10^{\frac{k-pK^+}{2}}, \tag{6.3}$$

we obtain

$$\left.\begin{array}{l} \alpha(k) = \dfrac{1}{a}\left\{\psi_0 - \left(7 - \dfrac{pK^+ + k}{2}\right)\ln 10 - \delta_0\right\}, \ \beta(x) = -pK^+ \\[2mm] -2\left(\dfrac{-ax + \psi_0}{\ln 10} - 7\right), \ \alpha_k(k) = \dfrac{\ln 10}{2a}, \ \beta_x = \dfrac{2a}{\ln 10}, \ \dfrac{2}{\ln 10}(\delta \\[2mm] -\psi_0) + 14 - pK^+ \leqslant k \leqslant \dfrac{2}{\ln 10}\{a + \delta_0 - \psi_0\} + 14 - pK^+. \end{array}\right\} \tag{6.4}$$

For the distribution $M(k)$ relative to the sorting parameter, we have [compare with (5.12)]

$$M(k) = -\frac{S\frac{\ln 10}{2a}\,\mathrm{sh}\left\{\left(7-\frac{pK^{+}+k}{2}\right)\ln 10 + \delta_0\right\}\left\{\mathrm{ch}\,\delta_0 + \frac{1}{2}\,10^{\frac{k-pK^{+}}{2}}\right\}}{\mathrm{sh}\,\delta_0},$$

$$\psi(x)\cdot\delta_0 < 0, \quad x \in [0,\,1].$$

(6.5)

Formulas (5.15)-(5.17) remain unchanged.

7. Creation of a Linear pH Profile (Third Model)

The models for creating a pH gradient with $\delta_0(k) = \mathrm{const}$ have an important disadvantage: they allow us to construct a pH gradient only for the values pH < 7 or pH > 7, without including the point pH $= 7$. We can avoid this if we do not assume that $\delta_0(k) \equiv \mathrm{const}$.

For concreteness, let us consider the model where we choose the quantity pI as the sorting parameter:

$$k \equiv pI. \tag{7.1}$$

Then [see (5.9)]

$$\left.\begin{array}{l}\psi^{i}(k) = (7-k)\ln 10 + \delta_0(k), \quad \psi^{e}(k) = (7-k)\ln 10, \\[2mm] \alpha(k) = \frac{1}{a}\{\psi_0 - (7-k)\ln 10 - \delta_0(k)\}.\end{array}\right\} \tag{7.2}$$

In order to determine the quantity $M(k)$, we have the expression

$$M(k) = -\frac{S\,|\,\alpha_k(k)|\,\mathrm{sh}\,\psi^{i}(k)\,\{\mathrm{ch}\,\delta_0(k) + m(k)\}}{\mathrm{sh}\,\delta_0(k)} \geqslant 0. \tag{7.3}$$

The created pH gradient will include the point pH $= 7$ if the quantity $\psi_i(k)$ changes sign (we recall that $K_w = 10^{-7}$ mole/liter).

Let

$$\psi^{i}(k_*) = 0. \tag{7.4}$$

Then, in order to satisfy the condition $M(k) \geq 0$, it is sufficient to require

$$\delta_0\,(k_*) = 0, \quad \delta_{0,k}\,(k) > 0, \quad k \in G. \tag{7.5}$$

It is easy to show that, in the case (7.2),

$$k_* = 7. \tag{7.6}$$

As an example, let us consider one of the simplest cases for the dependence $\delta_0(k)$; specifically,

$$\delta_0\,(k) = \delta_1\,(k - 7), \quad \delta_1 > 0, \quad \delta_1 = \text{const.}$$
$$\ln 10 > \delta_1 \tag{7.7}$$

In order to determine $M(k)$ and $E(x)$, we obtain

$$M\,(k) = -\frac{S\,\dfrac{\ln 10 - \delta_1}{a}\,\mathrm{sh}\,\{(7 - k)\,(\ln 10 - \delta_1)\}\,\left\{\mathrm{ch}\,\delta_1\,(k - 7) + \dfrac{1}{2}\,10^{G(k)}\right\}}{\mathrm{sh}\,\delta_1\,(k - 7)}, \tag{7.8}$$

$$E\,(x)\,\frac{j}{\mathrm{ch}\,(\psi\,(x) - \psi_*) + \dfrac{D^0\,(\beta\,(x))}{\mathrm{sh}\,\left\{\dfrac{\delta_1}{\ln 10 - \delta_1}\,\psi\,(x)\right\}}\,\mathrm{sh}\,\psi\,(x)}, \tag{7.9}$$

where

$$\beta\,(x) \equiv \frac{ax - \psi_0}{\ln 10 - \delta_1} + 7 \equiv -\frac{\psi\,(x)}{\ln 10 - \delta_1} + 7, \quad G = k - \mathrm{pK}^+\,(k). \tag{7.10}$$

8. Results of Calculations

In this section we present the results of calculations of the total concentrations $M(k)$ and the electric field intensity $E(x)$ using the three proposed models which allow us to create a linear pH profile. We should point out that some parameters that are essential for quantitative comparison with experiment are unknown at the present time. The parameters chosen for the calculations approximately correspond to those which are usually encountered in experimental practice. In addition, those parameters for which data is lacking [the diffusion coefficients $D^0(k)$ and the dependence of the difference between the isoelectric and isoionic points on the sorting parameter] were chosen so that the potential V and the current I approximately corresponded to the experimental values.

For all the calculations we used the following values of the parameters (in parentheses we give the corresponding dimensionless values) [12]: $S = 0.245 \cdot 10^{-3}$ m^2 ($0.245 \cdot 10^{-1}$), $V_0 = 0.245 \cdot 10^{-4}$ m^3, $D_{H_2O} =$

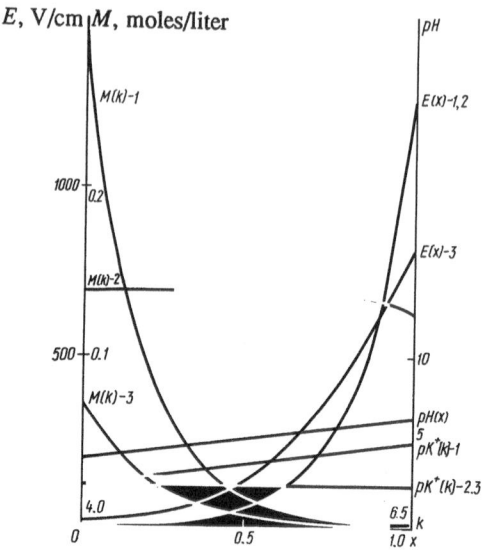

Fig. 47. Results of calculation of the total concentration $M(k)$ and the electric field intensity $E(x)$ using models 1, 2, and 3, which make it possible to create a specified linear pH profile ($4 \le$ pH ≤ 6.5).

TABLE 25. Results of Calculations Using the First Model ($4 \le$ pH ≤ 6.5)

$k = $ pl	E, V/cm	M, moles/ liter	σ, $1/\Omega \cdot m^*$	pK+	pK−
4.0	3.53	0.29139	0.17604	2.50	5.50
4.25	6.29	0.16386	0.09879	2.75	5.75
4.50	11.19	0.09214	0.05553	3.00	6.00
4.75	19.90	0.05182	0.03123	3.25	6.25
5.00	35.39	0.02914	0.01756	3.50	6.50
5.25	62.95	0.01638	0.00987	3.75	6.75
5.50	112.02	0.00921	0.00555	4.00	7.00
5.75	119.62	0.00517	0.00311	4.25	7.25
6.00	357.36	0.00288	0.00174	4.50	7.50
6.25	649.29	0.00159	0.00096	4.75	7.75
6.50	1239.79	0.00083	0.00050	5.00	8.00

$^* \sigma = \dfrac{i}{E}$ — conductivity of solution.

TABLE 26. Results of Calculations Using the Second Model (4 ≤ pH ≤ 6.5)

pI	E, V/cm	M, moles/liter	σ, 1/Ω·m	pK^+	$k =$ $= pK^-$
4.00	3.53	0.13703	0.17604	2.5	5.5
4.25	6.29	0.13703	0.09879	2.5	6.0
4.50	11.19	0.13703	0.05553	2.5	6.5
4.75	19.90	0.13702	0.03123	2.5	7.0
5.00	35.39	0.13701	0.01756	2.5	7.5
5.25	62.95	0.13698	0.00987	2.5	8.0
5.50	112.02	0.13689	0.00555	2.5	8.5
5.75	119.62	0.13659	0.00311	2.5	9.0
6.00	357.36	0.13565	0.00174	2.5	9.5
6.25	649.29	0.13267	0.00096	2.5	10.0

TABLE 27. Results of Calculations Using the Third Model (4 ≤ pH ≤ 6.5)

$k =$ pI	E, V/cm	M, moles/liter	σ, 1/Ω·m	pK^+	pK^-
4.00	13.26	0.07344	0.04688	2.50	5.50
4.25	21.72	0.04509	0.02861	2.75	5.75
4.50	35.31	0.02791	0.01759	3.00	6.00
4.75	56.82	0.01745	0.01093	3.25	6.25
5.00	90.32	0.01105	0.00688	3.50	6.50
5.25	141.34	0.00710	0.00439	3.75	6.75
5.50	216.78	0.00466	0.00287	4.00	7.00
5.75	323.72	0.00314	0.00192	4.25	7.25
6.00	466.25	0.00219	0.00133	4.50	7.50
6.25	639.01	0.00161	0.00097	4.75	7.75
6.50	818.92	0.00126	0.00076	5.00	8.00

TABLE 28. Results of Calculations Using the Third Model (3.5 ≤ pH ≤ 9.5)

$k =$ pI	E, V/cm	M, moles/liter	σ, 1/Ω·m	pK^+	pK^-
3.5	2.15	0.55075	0.79057	2.0	5.0
4.1	7.10	0.16693	0.23940	2.6	5.6
4.7	22.48	0.05285	0.07561	3.2	6.2
5.3	66.34	0.01795	0.02562	3.8	6.8
5.9	172.37	0.00693	0.00986	4.4	7.4
6.5	345.15	0.00346	0.00492	5.0	8.0
7.1	422.98	0.00283	0.00402	5.6	8.6
7.7	285.90	0.00419	0.00595	6.2	9.2
8.3	128.21	0.00932	0.01326	6.8	9.8
8.9	46.88	0.02546	0.03626	7.4	10.4
9.5	15.47	0.07707	0.10987	8.0	11.0

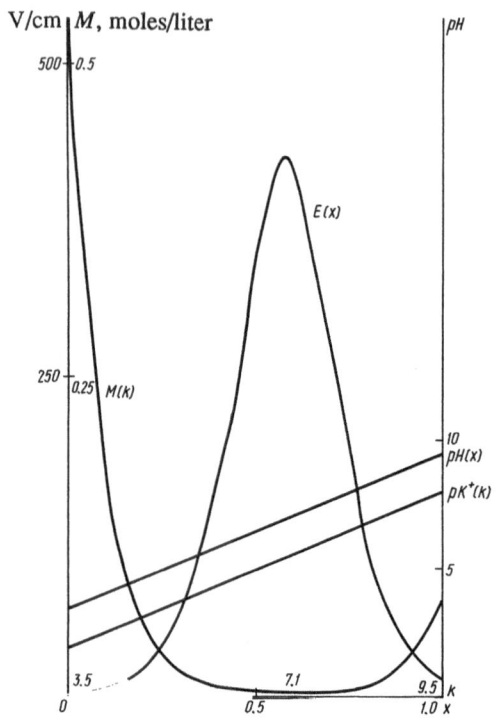

Fig. 48. Results of calculation of the total concentration $M(k)$ and the field intensity $E(x)$ using model 3 for the specified linear pH profile in the range $3.5 \leq pH \leq 9.5$.

$7.011191 \cdot 10^{-9}$ m^2/s, $D^0(k) = 1.067 \cdot 10^{-9}$ m^2/s (0.152215), $\psi_* = -0.2835813$, $M = 8.163265 \cdot 10^{-6}$ mole/liter, $K_w = 10^{-7}$ mole/liter, $L = 0.1$ m. Here, V_0 is the volume of the mixture; S, cross section of the considered volume; L, length of the volume; D_{H_2O}, diffusion coefficient for water; $D^0(k)$, diffusion coefficient for the components of the kth type; M, dimensional factor for going from $M(k)$ to the dimensional value.

1. **First model** ($4 < pH < 6.5$). $k = pI$, $4.5 \leq k \leq 6.5$, $G = pI - pK^+ = 1.5$, $\delta_0 = -2.302585 \cdot 10^{-3}$ (10^{-3} pH unit), $j = 62.14$ A/m^2 (1190.505), $V = 2000$ V (2), $I = 15.22 \cdot 10^{-3}$ A, $a = 5.756463$, $\psi(0) = 6.907755$ (pH = 4), $\psi(1) = 1.151292$ (pH = 6.5).

The results of the calculations are given in Table 25. In Fig. 47 we show the curves for $M(k)$ (1), $E(x)$ (1), pK$^+(k)$ (1), and the quantity pH(x).

TABLE 29. Set of Substances Suitable for Creating pH Gradients (from [46])

Carrier ampholyte	pI	pI − − pK+
Aspartic acid	2.77	0.89
Asparaglytyrosine	2.85	0.72
o-Aminophenyl-arsonic acid	3.00	0.77
Aspartylaspartic acid	3.04	0.34
p-Aminophenyl-arsonic acid	3.15	0.92
Picolinic acid	3.16	2.15
L-Glutamic acid	3.22	1.03
β-Hydroxyglutamic acid	3.29	0.96
Asparagyllysine	3.31	1.21
Isonicotinic acid	3.35	1.51
Nicotinic acid	3.44	1.37
Anthranilic acid	3.51	1.47
p-Aminobenzoic acid	3.62	1.30
Glycylaspartic acid	3.93	0.82
m-Aminobenzoic acid	4.29	0.81
Cystinyldiglycine	4.74	1.62
α-Hydroxyasparagine	4.92	2.43
α-Asparagylhistidine	4.94	1.90
β-Asparagylhistidine	4.96	2.00
Cysteinylcystine	5.32	2.31
Pentaglycine	5.40	2.27
Tetraglycine	5.59	2.35
Triglycine	5.60	2.33
Tyrosyltyrosine	5.85	2.08
Isoglutamine	6.10	2.04
Lysylglutamic acid	6.81	1.65
Histidylglycine	7.30	1.00
Histidylhistidine	7.47	0.50
Histidine	7.67	1.50
L-Methylhistidine	8.17	1.19
Carnosine	8.20	1.34
α,β-Diaminopropionic acid	8.27	1.40
Tyrosylarginine	8.68	1.00
L-Ornithine	9.70	1.05
Lysine	9.74	0.79
Lysyllysine	10.04	0.59
Arginine	10.76	1.72

TABLE 30. Results of Calculations for the Set of Substances Given in Table 29, Corresponding to Fig. 49

$k = pI$	x, mm	E, V/cm	M, moles/liter	σ, $1/\Omega \cdot m$	$pI - pK^+$
2.77	0	1.37	0.105487	0.74428	0.89
2.85	1.10	1.62	0.066410	0.63033	0.72
3.00	2.88	2.21	0.053106	0.46184	0.77
3.04	3.38	2.40	0.025974	0.42518	0.34
3.15	4.76	3.01	0.051107	0.33887	0.92
3.16	4.88	3.07	0.069525	0.33196	2.15
3.22	5.63	3.48	0.054609	0.29342	1.03
3.29	6.51	4.02	0.041423	0.25416	0.96
3.31	6.76	4.18	0.065163	0.24396	1.21
3.35	7.26	4.54	0.113322	0.22479	1.51
3.44	8.39	5.46	0.069946	0.18706	1.37
3.51	9.26	6.29	0.075228	0.16222	1.47
3.62	10.64	7.86	0.042015	0.12978	1.30
3.93	14.52	14.66	0.008891	0.06960	0.82
4.29	19.02	29.82	0.004322	0.03420	0.81
4.74	24.66	70.62	0.009510	0.01440	1.62
4.92	26.91	98.68	0.042390	0.01030	2.43
4.94	27.16	102.37	0.012759	0.00997	1.90
4.96	27.41	106.19	0.014829	0.00961	2.00
5.32	31.91	201.65	0.015896	0.00506	2.31
5.40	32.92	231.21	0.012673	0.00441	2.27
5.59	35.29	316.94	0.011138	0.00322	2.35
5.60	35.42	322.10	0.010472	0.00317	2.33
5.85	38.55	474.13	0.004049	0.00215	2.08
6.10	41.68	669.87	0.002630	0.00152	2.04
6.81	50.56	1234.97	0.000603	0.00083	1.65
7.30	56.69	1181.02	0.000162	0.00086	1.00
7.47	58.82	1057.87	0.000078	0.00096	0.50
7.67	61.33	881.68	0.000608	0.00116	1.50
8.17	67.58	465.00	0.000597	0.00219	1.19
8.20	67.96	444.72	0.000851	0.00229	1.34
8.27	68.84	399.92	0.001074	0.00255	1.40
8.68	73.97	204.59	0.000925	0.00499	1.00
9.70	86.73	31.11	0.006628	0.03280	1.05
9.74	87.23	28.78	0.004423	0.03546	0.79
10.04	90.99	15.96	0.005734	0.06392	0.59
10.76	100.00	3.74	0.224645	0.27294	1.72

2. **Second model** ($4 \leq pH \leq 6.5$). $k = pK^-, 5.5 \leq k \leq 10.5, pK^+ = 2.5$, $\delta_0 = -2.302585 \cdot 10^{-3}$ (10^{-3} pH unit), $j = 62.14$ A/cm^2 (1190.505), $V = 2000$ V (2), $I = 15.22 \cdot 10^{-3}$ A, $a = 5.756463$, $\psi(0) = 6.907755$ (pH = 4); $\psi(1) = 1.151292$ (pH = 6.5).

The results of the calculations are given in Table 26. In Fig. 47 we show the curves for $M(k)$ (2), $E(x)$ (2), $pK^+(k)$ (2), and the quantity $pH(x)$. We note that the field intensity is the same as in variant 1 (see Sections 5 and 6).

3. **Third model** ($4 \leq pH \leq 6.5$). $k = pI, 4 \leq k \leq 6.5, G = pI - pK^+ = 1.5, \delta_0(k) = \delta_1(k - 7), \delta_1 = 3.020616 \cdot 10^{-3}, j = 62.14$ A/cm^2

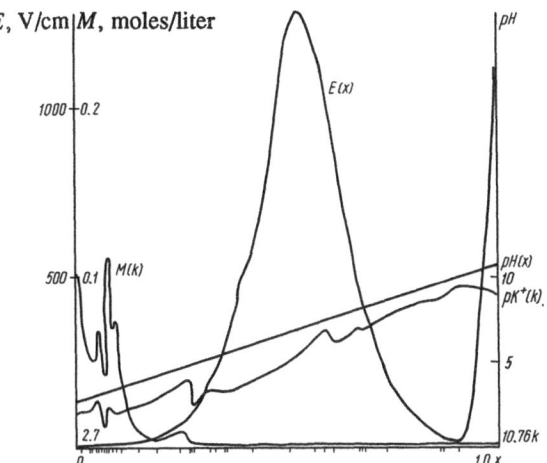

Fig. 49. Results of the calculation of the total concentration $M(k)$ and the field intensity $E(x)$ for the set of substances given in Table 29, using which a linear pH profile is created in the range $2.77 \leq pH \leq 10.76$.

(1190.505), $V = 2000$ V (2), $I = 15.22 \cdot 10^{-3}$ A, $a = 5.756463$, $\psi(0) = 6.907755$ (pH = 4); $\psi(1) = 1.161292$ (pH = 6.5).

The results of the calculations are given in Table 27. In Fig. 47 we show the curves for $M(k)$ (3), $E(x)$, $pK^+(k)$ (3), and the quantity $pH(x)$. The calculations for variants 1-3, which create the same pH profile, are carried out for the same value of VI. Variants 1 and 2 correspond to different mixtures (see the pK^-, pK^+ values in Tables 25 and 26). Variant 1 differs from variant 3 by the nature of the dependence $\delta_0(k)$. Comparison of the results of the calculations with experiments (qualitative comparison, of course), using, for example, the quantity $M(k)$, makes it possible to choose between the different models.

4. **Third model** $(3.5 \leq pH \leq 9.5)$. $k = pI$, $3.5 \leq k \leq 9.5$, $G = pI - pK^+ = 1.5$, $\delta_0(k) = \delta_1(k - 7)$, $\delta_1 = 4.60517 \cdot 10^{-4}$, $j = 169.974$ A/cm^2 (3256.294), $V = 1500$ V (1.5), $I = 41.64352 \cdot 10^{-3}$ A, $a = 13.8155$, $\psi(0) = 8.059047$ (pH = 3.5); $\psi(1) = -5.756462$ (pH = 9.5).

The results of the calculations are given in Table 28. In Fig. 48, we show the curves for $m(k)$, $E(x)$, $pK^+(k)$, and the quantity $pH(x)$. Calculations for this variant show that it is possible to create a linear pH profile including the point pH = 7.

5. **Third model** $(2.77 \leq pH \leq 10.76)$. $k = pI$, $2.77 \leq k \leq 10.76$, $G = pI - pK^+ \equiv G(k)$, $\delta_0(k) = \delta_1(k - 7)$, $\delta_1 = 2.302585 \cdot 10^{-3}$, $j = 102.04$ A/cm^2 (1954.861), $V = 3400$ V (3.4), $I = 25 \cdot 10^{-3}$ A, $a = 18.37925$, $\psi(0) = 9.730194$ (pH = 2.77); $\psi(1) = -8.649062$ (pH = 10.76).

In calculating this variant, we chose a specific set of substances which are suitable for creating the pH gradient. A list of these substances indicating the pI and pI $-$ pK$^+$ values is given in Table 29. The approximate form of the function $G(k)$ was plotted [in Fig. 49, we show the graph of the function $pK^+(k) \equiv pI - G(k) \equiv k - G(k)$]. The results presented demonstrate that it is possible in principle to calculate the concentrations of specific substances for creating a linear (and, generally speaking, any other) pH profile. Of course, the results obtained are qualitative and may be compared with experimental data only after refining the form of the functions $D^0(k)$ and $\delta_0(k)$. The form of these functions may be determined experimentally, by measuring the electrical conductivities (the molar electrical conductivities) and the electromigration rates for ions of the specific substances at different pH values.

Analysis of Tables 29 and 30 shows that significant changes in the total concentration $M(k)$ approximately coincide with sharp changes in the function $pK^+(k)$. This allows us to conclude that the function $pK^+(k)$ should be as smooth as possible in order to ensure a steady variation in the total concentration (this may be achieved if we exclude certain substances from the mixture). We also note that if the pH gradient is created using a finite, insufficiently large set of substances, the pH profile may differ from a linear profile, and the behavior of such a mixture will not be described by the infinite-component model. Nevertheless, the results of such calculations may help considerably in choosing the concentrations of specific substances in an N-component solution. For example, they may be chosen as the initial approximations and refined during the experiment.

9. Temperature Distribution in an Infinite-Component Mixture

As has been indicated, the thermal conductivity equation (2.4) is separated from the system and may be integrated if we know the form of the functions $E(x)$ and $Q(x)$. Let us consider the following problem:

$$- \varkappa T_{xx}(x) = \delta \{E(x) j - Q(x)\}, \tag{9.1}$$

$$T_0 = T(1) = T_0. \tag{9.2}$$

The boundary conditions correspond to specifying the temperature T_0 on the boundary. The function $Q(x)$ describes the work done by the external cooling device. Of course, withdrawal of heat with the aid of the cooling device occurs through the lateral boundaries of the solution. Rewriting Eq. (9.1), taking into account the heat "sink," corresponds to the case when the transverse heat flux is almost constant. This assumption is valid if the transverse dimension of the volume in which the solution is placed is significantly less than its longitudinal dimension, and temperature changes in the transverse direction may be neglected. The solution to problem (9.1), (9.2) has the form

$$T(x) = T_0 - \frac{\delta}{\varkappa} \int_0^x (x - s) \{E(s) j - Q(s)\} \, ds + \frac{\delta}{\varkappa} x \int_0^1 (1 - s) \{E(s) j - Q(s)\} \, ds. \tag{9.3}$$

The case where the capacity of the cooling device has the following form is of practical interest:

$$Q(x) = E(x) j. \tag{9.4}$$

In this case, the temperature distribution along the electrophoretic volume will be constant.

Let us now consider the two-dimensional case. Let us choose the volume occupied by the infinite-component mixture in the form of a cylinder of constant cross section $S = \pi r_0^2$, where r_0 is the radius of the cylinder. Let us introduce the cylindrical coordinates $\mathbf{x} = (r, x, \varphi)$, where r is the distance from the axis of the cylinder, x is the distance along the axis of the cylinder, and φ is the polar angle ($0 \leq r \leq r_0, 0 \leq x \leq 1, 0 \leq \varphi \leq 2\pi$). Let us assume that the concentration, the electric field intensity, and the acidity do not depend on r and φ, and the temperature does not depend on φ:

$$\xi(k, \mathbf{x}, t) = \xi(k, x, t), \ \mathbf{E}(\mathbf{x}, t) = E(x, t), \ \psi(\mathbf{x}, t), = \psi(x, t),\\ T(\mathbf{x}, t) = T(x, t), \ \mathbf{E}(x, t) = \{0, E(x, t), 0\}. \tag{9.5}$$

Then, in the steady-state case, system (1.3)-(1.6) is reduced to system (2.1)-(2.3), and the thermal conductivity equation takes on the form

$$- \varkappa \Delta T(r, x) = \delta \{E(x) j - Q(x)\}, \tag{9.6}$$

where the quantities $E(x), j$ are determined from the solution to problem (2.1)-(2.5).

Let us assume that there is no internal cooling device [$Q(x) = 0$], on the lateral surface of the cylinder the heat flux $p(x)$ is specified, and on the bases of the cylinder the temperature T_0 is specified. We note that, for the one-dimensional problem, the external cooling device is modeled using a volume heat sink. In the case of a two-dimensional problem, it is reasonable to model the external cooling device by specifying, for example, the heat flux at the boundary. In this case, we have the following boundary-value problem for determining the temperature:

$$\frac{1}{r}\frac{\partial}{\partial r}r\frac{\partial T}{\partial r} + \frac{\partial^2 T}{\partial x^2} = -\frac{\delta j}{\varkappa}E(x), \quad 0 \leqslant r \leqslant r_0, \quad 0 \leqslant x \leqslant 1 \quad (9.7)$$

$$-\varkappa\frac{\partial T}{\partial r}\Big|_{r=r_0} = P(x), \quad T(r, x)|_{r=0} < \infty, \quad 0 \leqslant x \leqslant 1, \quad (9.8)$$

$$T(r, x)|_{x=0} = T(r, x)|_{x=1} = T_0. \quad (9.9)$$

The problem obtained may be solved by the separation-of-variables method. Let us represent the solution in the form

$$T(r, x) = u(x) + w(r, x). \quad (9.10)$$

Let us determine the function $u(x)$ from the problem

$$u_{xx} = -\frac{\delta j}{\varkappa}E(x), \quad 0 \leqslant x \leqslant 1, \quad (\ )_x \equiv \frac{\partial}{\partial x}(\ ), \quad (9.11)$$

$$u(0) = u(1) = T_0, \quad (9.12)$$

solving which we obtain [compare with (9.1)-(9.3)]

$$u(x) = T_0 - \frac{\delta j}{\varkappa}\int_0^x (x-s)E(s)\,ds + \frac{\delta}{\varkappa}x\int_0^x (1-s)E(s)\,ds. \quad (9.13)$$

Substituting (9.10) into (9.7)-(9.9), and taking into account (9.11), (9.12), we derive

$$\frac{1}{r}(rw_r)_r + w_{xx} = 0, \quad (\ )_r \equiv \frac{\partial}{\partial r}(\ ), \quad (9.14)$$

$$-\varkappa w_r|_{r=r_0} = P(x), \quad w|_{r=0} < \infty, \quad (9.15)$$

$$w|_{x=0} = w|_{x=1} = 0. \quad (9.16)$$

Let us look for a solution to this problem in the form

$$w(r, x) = \sum_{n=1}^{\infty} a_n R_n(r) W_n(x). \quad (9.17)$$

In order to determine $R_n(r)$, $W_n(x)$, we have

$$W_{n,xx} = -\lambda_n^2 W_n, \quad W_n(0) = W_n(1) = 0, \tag{9.18}$$

$$\frac{1}{r}(rR_{n,r})_r = \lambda_n^2 R_n, \quad R_n(0) < \infty, \tag{9.19}$$

from which we have

$$W_n(x) \equiv \sin \pi n x, \quad \lambda_n = \pi n, \tag{9.20}$$

$$Rn(r) \equiv I_0(\pi n r), \tag{9.21}$$

where I_0 is the modified Bessel function. The set of functions $\{W_n\}_{n=1}^{\infty}$ is a fundamental sequence on the segment $[0, 1]$. Expanding $P(x)$ in a series in $W_n(x)$, we derive

$$P(x) = \sum_{k=1}^{\infty} \left\{ 2 \int_0^1 P(s) \sin \pi k s \, ds \right\} \sin \pi k x. \tag{9.22}$$

Substituting (9.17) into (9.15), and using (9.21), (9.22), we obtain

$$a_n = \frac{2}{\varkappa \pi n I_1(\pi n r_0)} \int_0^1 P(x) \sin \pi n x \, dx, \quad n = 1, 2, \ldots, \tag{9.23}$$

where I_1 is the modified Bessel function of the first kind ($I_0 = I_1$). Finally, for the temperature distribution, we have (also see [24])

$$T(r, x) = u(x) - \sum_{n=1}^{\infty} \frac{2\int_0^1 P(s)\sin \pi n s \, ds}{\varkappa \pi n I_1(\pi n r_0)} \sin \pi n x \cdot I_0(\pi n r). \tag{9.24}$$

In particular, if

$$P(x) \equiv P_0, \tag{9.25}$$

then

$$T(r, x) = u(x) - \sum_{m=0}^{\infty} \frac{4P_0 \sin(2m+1)\pi x}{\varkappa \pi^2 (2m+1)^2} \cdot \frac{I_0((2m+1)\pi r)}{I_1((2m+1)\pi r_0)} \cdot \tag{9.26}$$

Expanding in a series in r, we obtain

$$T(r, x) = u(x) - \Phi(x) + \frac{1}{4} \Phi_{xx}(x) r^2 + O(r^3), \quad r \to 0 \quad (9.27)$$

where

$$\Phi(x) \equiv \sum_{m=0}^{\infty} \frac{4P_0 \sin(2m+1)\pi x}{\varkappa \pi^2 (2m+1)^2 I_1 ((2m+1)\pi r_0)}. \quad (9.28)$$

In the general case, in order to maintain a constant temperature on the axis of the cylinder, it is necessary that the function $P(x)$ satisfy the relation

$$T_3 = u(x) - \sum_{n=1}^{\infty} \frac{2 \int_0^1 P(s) \sin \pi n s \, ds}{\varkappa \pi n I_1 (\pi n r_0)} \cdot \sin \pi n x, \quad (9.29)$$

where T_3 is the temperature on the axis of the cylinder (i.e., when $r = 0$).

Chapter VIII

RESOLUTION OF ISOELECTRIC FOCUSING

In this chapter, we study the nature of the behavior of amphoteric substances in a medium with a specified pH profile. We consider a multicomponent mixture in the case where we may neglect reaction between the components forming the pH gradient and the substances subject to isoelectric focusing, which, for the electrophoresis model used, inevitably leads to exclusion of reactions between the substances to be focused. This is connected with the fact that the reactions are taken into account in terms of the total charge of the mixture and the electric current density, in which are included the charge and the current of the component creating the pH gradient. In this case, it is impossible to separate the reactions between the individual substances to be focused and the reactions between these substances and the pH gradient.

The major goal of the treatment presented below is to rigorously formulate the idea of the resolution of isoelectric focusing and to analyze the effect of different factors on such resolution. The effect of thermal diffusion on the resolution of electrophoresis is thus apparently studied for the first time. In the literature [77, 93, 97], the question of the resolution is treated as part of the identification problem for nearby zones in the steady state. In particular, we may obtain dependences between the minimum difference of the isoelectric points for the two substances, the shape of the pH gradient, and the mobility function for the substances to be determined as a function of pH [93], when identification of individual zones is still possible. Such relations are also obtained in this chapter within the framework of the constructed mathematical model for electrophoresis.

In Section 1 of this chapter a mathematical model describing the behavior of the substances to be focused in a solution of carrier ampholytes is constructed by formally combining the equations of Chapter II for finite-

211

component mixtures and the equations of Chapter VII for infinite-component mixtures. In Section 2, for the one-dimensional case, we formulate the problem of describing the behavior of the part of the mixture to be investigated (or to be fractionated). This problem is the usual linear diffusion problem with transport, when the diffusion coefficients and the transport rates depend only on the location of the substances to be studied.

In Section 3 we investigate the steady state. We establish the principal terms of the asymptotic expansion of the solution for low diffusion ($\mu \to 0$). We introduce the idea of the resolution of the method and obtain the criteria for separation or identification of individual components of the mixture. We show that taking thermal diffusion into account as $\mu \to 0$ does not affect the resolution of the method, although it leads to a shift in the maximum concentration point. We note that an analogous effect may be caused by the dependence of the diffusion coefficient for a closed chemical subsystem on the pH of the medium. In Section 4, on the basis of the separation criterion formulated in Section 3, we obtain the condition for maximum difference between the isoelectric points of two components which may be separated for a specified pH profile.

The magnitude of the shift in the concentration maximum point under the influence of thermal diffusion for the specific model of amphoteric substances is determined in Section 5. We show that the magnitude of the shift is significantly less than the size of the interval in which the concentration of the given substance is localized in the steady state.

1. Formulation of the Problem

Let us assume that, in an infinite-component mixture consisting of closed chemical subsystems, there exists a finite number of chemical subsystems which cannot be characterized by a continuous sorting parameter. For the chemical subsystems which can be characterized by a continuous sorting parameter k, let the hypotheses made in Chapter VII be satisfied. The number of subsystems which cannot be described by a continuous sorting parameter is equal to N. Then the system of equations for describing the mixture has the form

$$\frac{\partial \xi (k,\ x,\ t)}{\partial t} + \operatorname{div} \mathbf{i} (k,\ x,\ t) = 0, \quad k \in G, \tag{1.1}$$

$$\mathbf{i} (k,\ x,\ t) = -\mu \nabla \{D\xi\} + \gamma \xi \mathbf{E}, \quad k \in G, \tag{1.2}$$

$$\frac{\partial \xi_m (x,\ t)}{\partial t} + \operatorname{div} \mathbf{i}_m (x,\ t) = 0, \quad m = 1,\ \ldots,\ N, \tag{1.3}$$

$$\mathbf{i}_m(\mathbf{x},\ t) = -\mu\nabla\left\{D_m(\psi)\,\xi_m(\mathbf{x},\ t)\right\} + \gamma_m(\psi)\,\xi_m(\mathbf{x},\ t)\,\mathbf{E}(\mathbf{x},\ t)$$
$$-\mu D_m^T\xi_m\nabla T, \tag{1.4}$$

$$\theta\operatorname{div}\mathbf{E} = \operatorname{sh}\psi + \int_G e\xi dk + \sum_{m=1}^{N} e_m(\psi)\,\xi_m, \tag{1.5}$$

$$\theta\frac{\partial E}{\partial t} + \mathbf{j} = J, \tag{1.6}$$

$$\mathbf{j} = \left\{-\mu\nabla\psi + \mathbf{E}\right\}\operatorname{ch}(\psi - \psi_*) + \int\left\{-\mu\nabla(\gamma\xi) + \sigma\xi\mathbf{E}\right\} dk$$

$$+ \sum_{m=1}^{N}\left\{-\mu\nabla(\gamma_m\xi_m) + \sigma_m\xi_m\mathbf{E} - \mu D_m^T\xi_m\nabla T\right\}, \tag{1.7}$$

$$\frac{\partial T}{\partial t} - \varkappa\Delta T = \delta\left\{\mathbf{E}\cdot\mathbf{j} - Q\right\}. \tag{1.8}$$

Let us consider the case where the concentrations ξ_m ($m = 1, ..., N$) of the subsystems which cannot be described by a continuous sorting parameter are low enough so that their contribution to the total electric current $\mathbf{j}$ and the free charge of the solution [i.e., to the right-hand side of Eq. (1.5)] may be neglected. Let us assume that the characteristic time for establishing a steady state for subsystems which can be described by a continuous sorting parameter is significantly shorter than the corresponding characteristic time for the remaining finite number of subsystems.

We note that if the part of the solution which can be described by a continuous sorting parameter is already found to be in a steady state, then the motion of the remaining part of the solution occurs in the specified electric field $\mathbf{E}(\mathbf{x})$. The mobilities, diffusion, and thermal diffusion coefficients, depending on the function $\psi = \psi(\mathbf{x})$, are also known functions of $\mathbf{x}$. The motion of the remaining part of the solution does not lead to a change in the functions $\mathbf{E}(\mathbf{x})$ and $\psi(\mathbf{x})$, since, by hypothesis, there is no reaction between parts of the solution (such reaction is accomplished in the given model through the free charge and the electric current).

Let a steady state be established for the part of the solution which can be described by a continuous sorting parameter. Then system (1.1)-(1.8) takes on the form

$$\operatorname{div}\mathbf{i}(k,\ \mathbf{x},\ t) \equiv \operatorname{div}\left\{-\mu\nabla\{D\xi\} + \gamma\xi\mathbf{E}\right\} = 0, \tag{1.9}$$

$$\theta\operatorname{div}\mathbf{E} = \operatorname{sh}\psi + \int_G e\xi dk, \tag{1.10}$$

$$\mathbf{j} = (-\mu\nabla\psi + \mathbf{E})\operatorname{ch}(\psi - \psi_*) + \int_G\left\{-\mu\nabla(\gamma\xi) + G\xi\mathbf{E}\right\} dk, \tag{1.11}$$

$$-\varkappa \Delta T = \delta (\mathbf{E} \cdot \mathbf{j} - Q), \tag{1.12}$$

$$\frac{\partial \xi_m}{\partial t} + \operatorname{div} \mathbf{i}_m = 0, \tag{1.13}$$

$$\mathbf{i}_m = -\mu \nabla \{D_m (\psi (x)) \xi_m\} + \gamma_m (\psi (x)) \xi_m \mathbf{E} (x) - \mu D_m^T (\psi (x)) \xi_m \nabla T (x). \tag{1.14}$$

2. Equations for Describing the Motion of the Mixture to be Separated in Isoelectric Focusing in a Specified pH Gradient

For simplicity, let us restrict ourselves to consideration of the one-dimensional case. Let conditions (VII.2.4)-(VII.2.5) be satisfied for Eqs. (1.9)-(1.12). Then the solution to problem (1.9)-(1.12), (VII.2.4)-(VII.2.5) describes the pH gradient created in the solution [the profile of the function $\psi(x)$], the intensity distribution $E(x)$, and the temperature distribution $T(x)$. As such a solution when $\theta = \theta_0 \mu$, $\mu \to 0$, we may choose the solution for one of the models proposed in Chapter VII. [We note that the functions $\psi(x)$, $E(x)$, $T(x)$ may be directly determined from experiment.] The motion of the components of the mixture to be separated or analyzed by isoelectric focusing (i.e., on the basis of the difference between the isoelectric points of the components) is described by Eqs. (1.13), (1.14), where $\psi(x)$, $E(x)$, $T(x)$ are known functions. For the boundary conditions, it is reasonable to choose the condition of the absence of fluxes at the boundary, i.e., impermeability of the boundary for the components of the mixture to be separated. Furthermore, we need to specify the concentration distribution in the mixture at the initial instant of time.

We note that, in practice, in preparative or analytical isoelectric focusing, an aqueous test solution of the mixture to be separated is transferred to a solution with a specified pH gradient. In this case, the acidity of the transferred solution may differ from the acidity of the carrier mixture. The considered model corresponds to the case where transfer of the test substance does not change the pH gradient profile; and in accordance with the assumptions concerning closed chemical subsystems and chemical reaction rates, rather rapid redistribution of the concentrations of the ions of the test substances occurs, corresponding to the value of the pH of the solution at which the separation will occur. The problem describing the behavior of the mixture to be separated has the form

$$\frac{\partial \xi_m (x, t)}{\partial t} + \frac{\partial i_m (x, t)}{\partial x} = 0, \quad m = 1, \ldots, N, \tag{2.1}$$

$$i_m(x, t) = -\mu \frac{\partial}{\partial x} \{D_m(\psi(x)) \xi_m\} + \gamma_m(\psi(x)) \xi_m E(x)$$

$$- \mu D_m^T(\psi(x)) \xi_m \frac{\partial T(x)}{\partial x}, \tag{2.2}$$

$$i_m(0, t) = i_m(1, t) = 0, \quad m = 1, \ldots, N, \tag{2.3}$$

$$\xi_m(x, 0) = \xi_m^{(0)}(x), \quad m = 1, \ldots, N. \tag{2.4}$$

The functions $D_m(\psi) > 0$, $\gamma_m(\psi)$, $D_m^T(\psi)$ have the form (II.3.12); the functions $\psi(x)$, $E(x)$ are assumed to be specified. It is easy to see that the motion of the individual components in the considered model is independent. Taking into account the effect of reactions would lead to the fact that the mixture to be separated would affect the variation in the functions $\psi(x)$, $E(x)$, since, as has already been indicated, reaction occurs through the free charge and electric current. Therefore, in the considered model, we cannot take into account reaction between the components of the mixture to be separated without taking into account reactions between the part of the mixture to be separated and the part of the mixture creating the pH gradient.

3. Basic Results

In the steady-state case, system (2.1)-(2.3) has the form

$$\frac{\partial i_m}{\partial x} \equiv \frac{\partial}{\partial x} \left\{ -\mu \frac{\partial}{\partial x}(D_m \xi_m) + \gamma_m \xi_m E - \mu D_m^T \xi_m \frac{\partial T}{\partial x} \right\} = 0, \tag{3.1}$$

$$i_m(0) = i_m(1) = 0, \quad m = 1, \ldots, N, \tag{3.2}$$

from which we get

$$-\mu \frac{\partial}{\partial x}(D_m \xi_m) + \gamma_m \xi_m E - \mu D_m^T \xi_m \frac{\partial T}{\partial x} = 0. \tag{3.3}$$

For the concentration we derive

$$\xi_m(x) = \xi_m(x_m) \frac{D_m(x_m)}{D_m(x)} \exp \left\{ -\int_{x_m}^{x} \frac{D_m^T(s) T_s(s)}{D_m(s)} ds \right\}$$

$$\times \exp \left\{ \frac{1}{\mu} \int_{x_m}^{x} \frac{\gamma_m(s) E(s)}{D_m(s)} ds \right\}. \tag{3.4}$$

Specifying the total concentration of the components as

$$M_m = S \int_0^1 \xi_m(x)\, dx = O(1), \quad m = 1, 2, \ldots, N, \qquad (3.5)$$

we determine

$$\xi_m(x_m) = \frac{M_m}{S} \left\{ \int_0^1 \left\{ \frac{D_m(x_m)}{D_m(x)} \exp\left\{ -\int_{x_m}^x \frac{D_m^T(s)\, T_s(s)}{D_m(s)}\, ds \right\} \exp \right. \right.$$

$$\left. \left. \times \left\{ \frac{\gamma_m(s)\, E(s)}{D_m(s)}\, ds \right\} dx \right\}^{-1}. \qquad (3.6)$$

Let us introduce the symbol

$$U_m(x) = \int_{x_m}^x \frac{\gamma_m(\psi(s))\, E(s)}{D_m(\psi(s))}\, ds. \qquad (3.7)$$

Let the following relations be satisfied (i.e., the point $x = x_m$ is the isoelectric point):

$$\gamma_m(\psi(x_m)) = 0, \quad \gamma_{m,\psi}(\psi) > 0,$$
$$\psi_x(x) < 0, \quad E(x) > 0, \quad 0 < x_m < 1. \qquad (3.8)$$

Then

$$U_{m,x}(x_m) = 0, \quad U_m(x_m) = 0, \quad U_m(x < 0, \quad x \neq x_m, \qquad (3.9)$$

$$U_{m,xx}(x_m) = \frac{\gamma_{m,\psi}(\psi(x_m))\, \psi_x(x_m)\, E(x_m)}{D_m(\psi(x_m))} < 0. \qquad (3.10)$$

Setting $\mu \to 0$, we obtain from (3.6), using the Laplace method (compare with Section 3 in Chapter VII),

$$\xi_m(x_m) = \frac{M_m}{S} \sqrt{-\frac{U_{m,xx}(x_m)}{2\pi\mu}} + O(\sqrt{\mu}), \quad \mu \to 0. \qquad (3.11)$$

Let us determine the characteristic width of the zone for the mth component by the relationship

$$\eta_m M_m = S \int_{x_m - \varepsilon_m}^{x_m + \varepsilon_m} \xi_m(x)\, dx, \quad m = 1, 2, \ldots, N,$$

$$(3.12)$$

$$0 < \eta_m < 1.$$

Here, η_m is the relative (compared with the total) amount of the mth component on the segment $[x_m - \varepsilon_m, x_m + \varepsilon_m]$. Using (3.5), we may rewrite this relationship in the form

$$(1 - \eta_m)\, M_m = S \left\{ \int_0^{x_m - \varepsilon_m} \xi_m(x)\, dx + \int_{x_m + \varepsilon_m}^1 \xi_m(x)\, dx \right\},$$

$$0 < x_m - \varepsilon_m < x_m + \varepsilon_m < 1. \tag{3.13}$$

Let

$$\varepsilon_m = O\left(\mu^{\frac{1+\alpha}{3}}\right), \quad 0 < \alpha < \frac{1}{2}. \tag{3.14}$$

Applying the Laplace method [53], from (3.11), (3.13) we derive

$$\frac{(1 - \eta_m)\, M_m}{S}$$

$$= \mu \, \frac{\{f(x_m) - f_x(x_m)\,\varepsilon_m + O(\varepsilon_m^2) + O(\mu)\}\, e^{\frac{1}{2\mu} U_{m,xx}(x_m)\varepsilon_m^2} (1 + O(\mu^\alpha))}{-U_{m,xx}(x_m)\,\varepsilon_m + O(\varepsilon_m^2)}$$

$$- \mu \, \frac{\{f(x_m) + f_x(x_m)\,\varepsilon_m + O(\varepsilon_m^2) + O(\mu)\}\, e^{\frac{1}{2}U_{m,xx}(x_m)\varepsilon_m^2} (1 + O(\mu^\alpha))}{U_{m,xx}(x_m)\,\varepsilon_m + O(\varepsilon_m^2)} \tag{3.15}$$

or, to an accuracy up to terms of order $O(\mu^\alpha)$,

$$\frac{1}{\sqrt{\pi}} \, \frac{e^{\frac{1}{2\mu}\,U_{m,xx}(x_m)\,\varepsilon_m^2}}{\sqrt{-\dfrac{U_{m,xx}(x_m)\,\varepsilon_m^2}{2\mu}}} = (1 - \eta_m), \tag{3.16}$$

where

$$f_m(x) \equiv \xi_m(x_m)\, \frac{D_m(x_m)}{D_m(x)} \exp\left\{-\int_{x_m}^x \frac{D_m^T(s)\, T_s(s)}{D_m(s)}\, ds\right\}. \tag{3.17}$$

In deriving (3.15), (3.17), we have used the relations

$$
\left.
\begin{aligned}
f_m(x_m \pm \varepsilon_m) &= f_m(x_m) \pm f_{m.x}(x_m)\,\varepsilon_m + O(\varepsilon_m^2),\\
U_m(x_m \pm \varepsilon_m) &= \tfrac{1}{2} U_{m,xx}(x_m)\,\varepsilon_m^2 + O(\varepsilon_m^3),\\
U_{m,x}(x_m \pm \varepsilon_m) &= \pm U_{m,xx}(x_m)\,\varepsilon_m + O(\varepsilon_m^2).
\end{aligned}
\right\} \tag{3.18}
$$

Let

$$\varepsilon_m = O\left(\mu^{\frac{1-\beta}{3}}\right), \quad 0 < \beta < \frac{1}{3}. \tag{3.19}$$

Then, to an accuracy up to terms of order $O(\mu^\beta)$, expression (3.16) may be replaced by the following:

$$Erf\left\{-\frac{U_{m,xx}(x_m))\,\varepsilon_m^2}{2\mu}\right\} = \eta_m, \quad Erf\,x \equiv \frac{2}{\sqrt{\pi}}\int_0^x e^{-s^2}ds, \tag{3.20}$$

from which, using (3.10), we obtain

$$\varepsilon_m = \sqrt{-\frac{2\mu D_m\,(\psi\,(x_m))}{\gamma_{m,\psi}\,(\psi\,(x_m))\,\psi_x\,(x_m)\,E\,(x_m)}}\;\Phi\,(\eta_m). \tag{3.21}$$

Here, Φ is the function inverse to Erf.

Let us say that for specified $\eta = \eta_m = \eta_{m+1}$ the components m and $m + 1$ are separated if the following relation is satisfied:

$$0 < \varepsilon_m + \varepsilon_{m+1} < x_{m+1} - x_m. \tag{3.22}$$

We will call the quantity $\varepsilon = \varepsilon_m + \varepsilon_{m+1}$ the separation resolution.

From the relations obtained it follows that, to an accuracy up to terms of order $\max\{O(\mu^\beta), O(\mu^\alpha)\}$, thermal diffusion (in the considered one-dimensional model) does not affect the separation resolution. However, we should point out that the presence of thermal diffusion leads to a shift in the concentration maximum for the mth component relative to the isoelectric point. Let us demonstrate this by rewriting expression (3.4) using (3.7), (3.17) in the form

$$\xi_m\,(x) = f_m\,(x)\,e^{\frac{1}{\mu}U_m(x)}, \tag{3.23}$$

from which we get

$$\xi_{m,x}\,(x) \equiv \left\{f_{m,x}\,(x) + \frac{f_m}{\mu}\,U_{m,x}\,(x)\right\}e^{\frac{1}{\mu}U_m(x)}$$

$$= \left\{f_{m,x}\,(x) + \frac{f_m\,(x)}{\mu}\,\frac{\gamma_m\,(\psi\,(x))\,E\,(x)}{D_m\,(\psi\,(x))}\right\}e^{\frac{1}{\mu}U_m(x)}. \tag{3.24}$$

It is easy to see that

$$\xi_{m,x}(x_m) = f_{m,x}(x_m)\, e^{\frac{1}{\mu}U_m(x_m)} = -\xi_m(x_m)\left\{ \frac{D_{m,\psi}(\psi(x_m))\,\psi_x(x_m)}{D_m(\psi(x_m))} \right.$$
$$\left. + \frac{D_m^T(\psi(x_m))\,T_x(x_m)}{D_m(\psi(x_m))} \right\}. \tag{3.25}$$

From this expression it is evident that, in the general case,

$$\xi_{m,x}(x_m) \neq 0,\ \max_{0 \leqslant x \leqslant 1} \xi_m(x) \neq \xi_m(x_m). \tag{3.26}$$

Let us consider several special cases. Let

$$D_{m,\psi} \equiv 0,\ D_m > 0,\ D_m^T(\psi(x_m))\,T_x(x_m) \neq 0. \tag{3.27}$$

In this case, the shift in the maximum occurs to the left of the point x_m if $D_m^T T_x > 0$, and to the right of the point x_m if $D_m^T T_x < 0$. Now let

$$D_{m,\psi}(\psi(x_m)) \neq 0,\quad \psi_x(x_m) \neq 0,\quad D_m^T(\psi(x_m))\,T_x(x_m) = 0. \tag{3.28}$$

In this case, the shift in the maximum point is connected with the dependence of the diffusion coefficient on acidity.

The shift in the maximum point for the mth component will be calculated in the next sections for specific models of the mixtures to be separated.

4. Conditions Imposed on the Difference Between Isoelectric Points for Components to be Separated

Let us consider the case where the mth and $(m + 1)$-th components have different isoelectric points. Let us introduce the symbol

$$\Delta pI = (\psi_m^i - \psi_{m+1}^i) \cdot \ln 10 > 0 \quad \begin{pmatrix} \gamma_{m+1}(\psi_{m+1}^i) \equiv 0 \\ \gamma_m(\psi_m^i) \equiv 0 \end{pmatrix}. \tag{4.1}$$

Here ψ_{m+1}^i, ψ_m^i are the values of the acidity for the isoelectric points; ΔpI is the difference between the isoelectric points in pH units. In the steady-state case, the following relations are satisfied:

$$\psi(x_{m+1}) = \psi_{m+1}^i,\ \psi(x_m) = \psi_m^i. \tag{4.2}$$

In this case,

$$\psi(x_{m+1}) = \psi(x_m) + \psi_x(x_m)(x_{m+1} - x_m) + O((x_{m+1} - x_m)^2),$$

$$|x_{m+1} - x_m| \to 0. \tag{4.3}$$

To an accuracy up to terms on the order of $O((x_{m+1} - x_m)^2)$, the following equality is satisfied:

$$x_{m+1} - x_m = \frac{\psi(x_{m+1}) - \psi(x_m)}{\psi_x(x_m)}, \quad \psi_x(x_m) < 0. \tag{4.4}$$

Specifying $\eta = \eta_m = \eta_{m+1}$, for concreteness let us set

$$\varepsilon_m = \min(\varepsilon_m, \varepsilon_{m+1}). \tag{4.5}$$

Then from (3.22), to an accuracy up to terms of order $O((x_{m+1} - x_m)^2)$, it follows that

$$2\Phi(\eta)\sqrt{-\frac{2\mu D_m(\psi_m')}{\gamma_{m,\psi}(\psi_m')\psi_x(x_m)E(x_m)}} \leqslant \frac{\psi_m^i - \psi_{m+1}^i}{-\psi_x(x_m)}, \tag{4.6}$$

from which we get

$$2\sqrt{2}\,\Phi(\eta)\sqrt{-\frac{\mu D_m(\psi_m')\psi_x(x_m)}{\gamma_{m,\psi}(\psi_m')E(x_m)}} \leqslant \psi_m^i - \psi_{m+1}^i \tag{4.7}$$

or, in pH units,

$$2\sqrt{2}\,\Phi(\eta)\sqrt{\frac{\mu D_m \dfrac{dpH}{dx}}{-E(x_m)\dfrac{d\gamma_m}{dpH}}} \leqslant \Delta pI. \tag{4.8}$$

The minimum value of ΔpI for which separation of the mth and $(m + 1)$-th components is possible is (for a specified η)

$$\Delta pI_{\min} = 2\sqrt{2}\,\Phi(\eta)\sqrt{\frac{\mu D_m \dfrac{dpH}{dx}}{-E(x_m)\dfrac{d\gamma_m}{dpH}}}, \tag{4.9}$$

Usually we choose (see, for example [77]),

$$2\sqrt{2}\,\Phi\,(\eta) = 3.07, \quad \eta = \mathit{Erf}\left(\frac{3.07}{2\sqrt{2}}\right) \approx 0.8752065. \tag{4.10}$$

5. Shift in the Concentration Maximum Point Due to Thermal Diffusion

Let us consider an N-component mixture to be separated for which the diffusion and thermal diffusion coefficients, the mobilities, and the molar charges have the form (II.3.12). Close to the isoelectric (and isoionic) point, usually $\beta_m{}^j = 0$ $(j \neq 1, j \neq -1, j \neq 0)$. Then

$$\left.\begin{aligned}
&D_m\,(\psi) = D_m^1\beta_m^1 + D_m^{-1}\beta_m^{-1} + D_m^0\,(1 - \beta_m^1 - \beta_m^{-1}), \quad \gamma_m\,(\psi) \\[4pt]
&= D_m^1\beta_m^1 - D_m^{-1}\beta_m^{-1}, \quad e_m\,(\psi) = \beta_m^1 - \beta_m^{-1}, \quad \beta_m^1 = \frac{\frac{1}{2}e^{\frac{\psi-\psi_m^e}{}}}{\text{ch}\,(\psi - \psi_m^e) + m_m}, \\[6pt]
&\beta_m^{-1} = \frac{\frac{1}{2}e^{-(\psi-\psi_m^e)}}{\text{ch}\,(\psi - \psi_m^e) + m_m}, \quad \psi_m^1 - \psi_m^e = \frac{1}{2}\ln\frac{D_m^{-1}}{D_m^1}, \quad m_m \\[6pt]
&= \frac{1}{2}\sqrt{\frac{K_m^+}{K_m^-}}, \quad \psi_m^e = \ln 10^7\sqrt{K_m^+ + K_m^-}, \quad D_m^T\,(\psi) \\[6pt]
&= D_m^0 D_m^{T,0}\,(1 - \beta_m^1 - \beta_m^{-1}) + D_m^{-1}D_m^{T,-1}\beta_m^{-1} + D_m^1 D_m^{T,1}\beta_m^1.
\end{aligned}\right\} \tag{5.1}$$

These relations may be rewritten in the form

$$\left.\begin{aligned}
&\gamma_m\,(\psi) = \sqrt{D_m^{-1}D_m^1}\,\frac{\text{sh}\,(\psi - \psi_m^i)}{\text{ch}\,(\psi - \psi_m^e) + m_m}, \quad e_m\,(\psi) = \frac{\text{sh}\,(\psi - \psi_m^e)}{\text{ch}\,(\psi - \psi_m^e) + m_m}, \\[6pt]
&D_m\,(\psi) = \sqrt{D_m^{-1}D_m^1}\,\frac{\text{ch}\,(\psi - \psi_m^i)}{\text{ch}\,(\psi - \psi_m^e) + m_m} + D_m^0\,\frac{m_m}{\text{ch}\,(\psi - \psi_m^e) + m_m}.
\end{aligned}\right\} \tag{5.2}$$

Let us assume that

$$D_m^0 = \sqrt{D_m^{-1}D_m^1}, \quad \psi_m^i = \psi_m^e, \quad D_m^{T,0} = D_m^{T,1} = D_m^{T,-1}. \tag{5.3}$$

Then

$$D_m\,(\psi) \equiv D_m^0; \quad \gamma_m\,(\psi) = D_m^0 e_m\,(\psi) = \frac{\text{sh}\,(\psi - \psi_m^e)}{\text{ch}\,(\psi - \psi_m^e) + m_m} \cdot D_m^0,$$
$$D_m^T\,(\psi) = D_m^{T,0}D_m^0. \tag{5.4}$$

Substituting the expressions obtained into (3.24), we derive

$$\xi_{m,x}(x) = \xi_m(x) \left\{ \frac{e_m(x) E(x)}{\mu} - D_m^{T,0} T_x(x) \right\}. \tag{5.5}$$

Let the point $x = y$ be the maximum point for the function $\xi_m(x)$. Then

$$e_m(y) E(y) - \mu D_m^{T,0} T_x(y) = 0. \tag{5.6}$$

We will look for roots to this equation in the form

$$y = x_m + \mu y_1 + \mu^2 y_2 + \dots, \quad \mu \to 0. \tag{5.7}$$

Substituting (5.7) into (5.6), expanding in a series about $x = x_m$, and equating terms with the same powers of μ, we obtain expressions for y_p ($p = 1, 2, \dots$). Thus, for y_1 we have

$$y_1 = \frac{D_m^{T,0} T_x(x_m)}{E(x_m) e_{m,\psi}(x_m) \psi_x(x_m)}. \tag{5.8}$$

Thus, for the isoelectric focusing model used, the shift in the concentration maximum point for the component is a quantity on the order of $O(\mu)$, while the concentration is mainly localized in an interval on the order of max $\{O(\mu^\beta), O(\mu^\alpha)\}$, where $0 < \beta < 1/3$, $0 < \alpha < 1/2$. This allows us to conclude that the shift in the concentration maximum point for the component as a result of thermal diffusion does not decrease the resolution of the method.

Chapter IX

ZONE EVOLUTION
IN ISOELECTRIC FOCUSING

The isoelectric focusing process is clearly subdivided into two qualitatively different stages (see Chapter III). Directly after the electric field is turned on, the charged particles existing in the solution begin to move. Assuming that the hypothesis of local chemical equilibrium is valid, this process of migration in a field may be described as the motion of quasisubstances, which in Chapter II are called chemical subsystems and can be characterized by individual electrophoretic mobilities. Let us recall that there are two types of chemical systems: "buffers," whose mobilities do not change sign in the electrophoretic space; and "samples," moving to their own isoelectric point, at which their mobility changes sign.

In the first stage of the process, motion of the primary mass of the sample occurs far from its isoelectric point, and the effect of diffusion is negligibly small. If in general there were no diffusion, then narrowing of the zone to be focused would continue without limit and the width of the zone would tend toward zero over the course of time. However, when the zone width close to the isoelectric point becomes comparable with the diffusion length, thermodynamic fluctuations interfere with this process. In this second stage, the concentration profile for the sample and the zone width are formed by simultaneous action of the electric field and diffusion. As a result of this process, a steady-state concentration profile is established for the sample. Since the diffusion coefficients are usually small, this stage of the process occurs considerably more slowly than the first stage.

In this chapter we consider a mathematical model for isoelectric focusing in the case where the sample does not affect the properties of the buffer solution (a strong buffer). We investigate both stages of isoelectric focusing on the basis of the asymptotic theory for low diffusion. The asymptotic solution obtained in particular allows us to observe the asymmetry of the focused zone, which causes an asymmetry in the characteristics of the sample relative to its isoelectric value. An unexpected conclusion is that the minimum zone width is not always achieved under steady-state conditions, so that an increase in the duration of the process does not always lead to an increase in its efficiency.

1. One-Dimensional Case

Let us consider an electrophoretic column whose length is significantly greater than its inner diameter. In this case, we may assume that the concentrations of buffer and sample, and also the electric field intensity distribution, do not depend on the distance from the axis of the cylinder. Then the isoelectric focusing problem (III.5.15)-(III.5.20) is reduced to the one-dimensional problem:

$$\frac{\partial \xi_1}{\partial t} + \frac{\partial}{\partial x} \left\{ - \mu D_4 \frac{\partial \xi_1}{\partial x} + \beta e^{-\psi} \text{sh} (\psi - \psi^i) \xi_1 \right\} = 0, \qquad (1.1)$$

$$\frac{\partial \xi_2}{\partial t} - \mu D_2 \frac{\partial^2 \xi_2}{\partial x^2} = 0, \qquad (1.2)$$

$$e^{\psi} = \sqrt{\xi_2}, \quad E = \frac{1}{\sqrt{\xi_2} (1 + D_3)} j(t), \quad \beta = \frac{Dj(t)}{m(1 + D_3)}, \qquad (1.3)$$

$$\xi_1 |_{t=0} = p(x), \quad \xi_2 |_{t=0} = \tilde{\xi}_2(x), \qquad (1.4)$$

$$i_{\xi_1} \equiv \left\{ - \mu D_4 \frac{\partial \xi_1}{\partial x} + \beta e^{-\psi} \text{sh} (\psi - \psi^i) \xi_1 \right\} = 0, \quad x = 0, 1, \qquad (1.5)$$

$$\xi_2 (0, \ t) = \xi_2^{(0)}(t), \quad \xi_2 (1, \ t) = \xi_2^{(1)}(t). \qquad (1.6)$$

Conditions (1.4) correspond to specification of the initial distribution of sample and buffer along the length of the electrophoretic column. Relation (1.5) is the condition of impermeability of the boundary of the column ($x = 0, x = 1$) for the sample. Finally, conditions (1.6) correspond to specification of the buffer concentration at the boundary of the column. The functions $p(x)$, $\tilde{\xi}_2(x)$, $\xi_1(t)$, $\xi_2^1(t)$ are assumed to be known.

Our goal is to study the motion of the sample in the vicinity of the isoelectric point, at which the acidity ψ of the mixture coincides with the isoelectric point of the sample ψ^i. Let us specify the initial concentration distribution of the buffer in such a way that ψ^i is included in the interval of variation for the quantity $\psi(x, t)|_{t=0}$. In the considered model, the concentration distribution for the buffer $\xi_2(x, t)$ and, consequently, the acidity distribution $\psi(x, t)$ and the electric field intensity distribution $E(x, t)$ do not depend on the sample concentration $\xi_1(x, t)$. Thus, the motion of the sample in the mixture does not affect its basic properties. Mathematically, this means that Eqs. (1.2), (1.3) with conditions (1.4), (1.6) may be solved independently of Eq. (1.1). In the case where $\xi_2^{(1)}(t)$, $\xi_2^{(0)}(t)$ do not depend on time and the initial distribution $\widetilde{\xi}_2(x)$ is a linear function, the solution has the form

$$
\left.
\begin{aligned}
&\xi_2(x,\ t) \equiv \xi_2(x) \equiv Ax + B,\ A = \xi_2^{(1)} - \xi_2^{(0)} = \text{const},\ B = \xi_2^{(0)} \\
&= \text{const},\ \psi(x, t) = \frac{1}{2} \ln(Ax + B),\ E(x,\ t) = \frac{j(t)}{\sqrt{Ax + B}\,(1 + D_3)}\ .
\end{aligned}
\right\} \quad (1.7)
$$

In the general case, we will assume that the quantity $\psi(x, t)$ is a known function determined for the solution to system (1.2), (1.3) with conditions (1.4), (1.6). Let us introduce the symbols

$$
\mu_0 = \mu D_4,\ f(x,\ t) = \beta e^{-\psi(x, t)} \operatorname{sh}(\psi(x,\ t) - \psi^i). \quad (1.8)
$$

Then the problem of determining the motion of the sample is reduced to the equation (we omit the index 1 for convenience: $\xi_1 \equiv \xi$)

$$
\frac{\partial \xi}{\partial t} + \frac{\partial}{\partial x}\left\{ -\mu_0 \frac{\partial \xi}{\partial x} + f(x,\ t)\xi \right\} = 0 \quad (1.9)
$$

with initial condition

$$
\xi(x,\ 0) = p(x) \quad (1.10)
$$

and boundary conditions

$$
\left\{ -\mu_0 \frac{\partial \xi}{\partial x} + f(x,\ t)\xi \right\}\Big|_{x=0,1} = 0. \quad (1.11)
$$

Let us determine the isoelectric points $x_0(t)$ in the mixture using the relations

$$
f(x_0(t),\ t) = 0,\ (\psi(x_0(t),\ t) \equiv \psi^i). \quad (1.12)
$$

In the following, we will assume that

$$
0 < x_0(t) < 1,\ t > 0. \quad (1.13)
$$

Violation of this condition means that isoelectric focusing under the given conditions cannot be accomplished within the electrophoretic column.

We note that in the case where $\psi(x, t)$ has the form (1.7), the value $x_0(t)$ does not depend on time and is

$$x_0 = \frac{e^{2\psi^t} - \xi_2^{(0)}}{\xi_2^{(1)} - \xi_2^{(0)}}. \tag{1.14}$$

Condition (1.13) imposes restrictions on the quantities $\xi_2^{(0)}$ and $\xi_2^{(1)}$ for a specified ψ^i.

2. Solution in the Case of Vanishing Diffusion

Let us assume that diffusion in the mixture is low:

$$\mu \to 0. \tag{2.1}$$

Then to a sufficient degree of accuracy for times significantly shorter than the characteristic diffusion time for the buffer, we may assume

$$\psi(x, t) \equiv \psi(x), \quad x_0(t) \equiv x_0, \quad f(x, t) \equiv f(x), \quad f(x_0) = 0. \tag{2.2}$$

In the following, we assume that condition (2.1) is satisfied. Let us consider problem (1.9)-(1.11), setting

$$\mu_0 = 0. \tag{2.3}$$

Using the method of characteristics (see, for example, [40, 45]), we rewrite Eq. (1.9) in the form

$$\frac{dx}{dt} = f(x), \tag{2.4}$$

$$\frac{d\xi}{dt} = -f_x(x)\,\xi, \tag{2.5}$$

where

$$\frac{d(\ )}{dt} \equiv \frac{\partial(\ )}{\partial t} + f(x)\,\frac{\partial(\ )}{\partial x}. \tag{2.6}$$

On the straight line $(x, 0)$ on the (x, t) plane, let us choose the point $a \neq x_0$ – the intersection between the characteristic $x(t)$ and the line $(x, 0)$. Then the initial conditions for problem (2.4), (2.5) have the form

$$x(0) = a, \quad \xi(0) = p(a). \tag{2.7}$$

Integrating (2.4), (2.5) and taking into account (2.7), we obtain

$$\xi(x, t) = p(a(x, t)) \frac{f(a(x, t))}{f(x)}. \tag{2.8}$$

We determine the function $a(x, t)$ by solving the equation of the characteristics relative to a:

$$\int_a^x \frac{ds}{f(s)} = t. \tag{2.9}$$

Let us require that the following condition be satisfied:

$$f_x(x) < 0, \quad x \in [0, 1]. \tag{2.10}$$

In this case, the velocity of all the particles of the sample found to the left of the point x_0 is directed to the right, and the velocity for all particles found to the right of the point x_0 is directed to the left:

$$\frac{dx}{dt} > 0, \ x < x_0; \quad \frac{dx}{dt} < 0, \ x > x_0; \quad \frac{dx}{dt} = 0, \ x = x_0. \tag{2.11}$$

Furthermore, if $p(a) > 0$ (obviously, the concentration is a positive quantity), then

$$\frac{d\xi}{dt} > 0, \ \xi(x, t) > 0. \tag{2.12}$$

Thus, condition (2.10) is a sufficient condition for the point x_0 to be a focusing point for the sample.

Let us note that the characteristics starting from the point $a \neq x_0$, i.e., the solutions to Eq. (2.4) with conditions (2.7), do not intersect the line (x_0, t) on the (x, t) plane. In the opposite case, from (2.8) it follows that $\xi(x, t)$ changes sign.

Analogously, for the characteristics starting from the point t_* of the line $(0, t)$ and the point t_{**} of the line $(1, t)$, specifying the conditions

$$x(t_*) = 0, \ \xi(t_*) = 0; \quad x(t_{**}) = 1, \ \xi(t_{**}) = 0, \tag{2.13}$$

we obtain

$$\xi(x, t) \equiv 0, \ x < x_1(t), \ x > x_2(t). \tag{2.14}$$

The functions $x_1(t)$ and $x_2(t)$ are determined by the equations [the characteristics starting from the points $(0, 0)$ and $(1, 0)$]

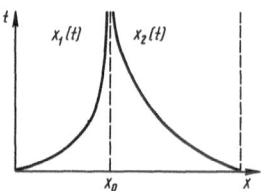

Fig. 50. Functions $x_1(t)$, $x_2(t)$ which bound the sample focused at the point x_0.

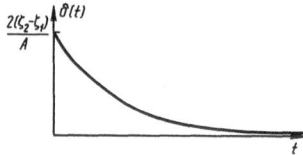

Fig. 51. Variation over time of the width of the focused zone.

$$\int_0^{x_1(t)} \frac{ds}{f(s)} = t, \quad \int_1^{x_2(t)} \frac{ds}{f(s)} = t. \tag{2.15}$$

Finally, for the characteristic starting from the point $a = x_0$, we have

$$x(0) = x_0, \quad \xi(0) = p(x_0). \tag{2.16}$$

From (2.4), (2.5) we obtain

$$x(t) = x_0, \quad \xi(x, t) = p(x_0) e^{-f_x(x_0) \cdot t}. \tag{2.17}$$

Thus, for any fixed instant of time t for vanishing diffusion ($\mu = 0$), the sample is localized on the interval

$$[x_1(t), x_2(t)] \quad (x_0 \in [x_1(t), x_2(t)]). \tag{2.18}$$

Let us give some results for the case where a linear pH profile is specified:

$$\psi(x) = Ax + B. \tag{2.19}$$

Then

$$f(x) = \beta e^{-\psi(x)} \operatorname{sh}[\psi(x) - \psi^i] \equiv \beta e^{-Ax - B} \operatorname{sh}[Ax + B - \psi^i]. \tag{2.20}$$

$$x_0 = \frac{\psi^i - B}{A}.$$

In order for conditions (1.13), (2.10) to be satisfied, we require that the following relationships be satisfied:

$$A\beta > 0, \quad \frac{\psi^i - B - A}{A} < 0, \quad \frac{\psi^i - B}{A} > 0. \tag{2.21}$$

Substituting (2.20) into (2.15), we derive

$$x_1(t) = x_0 + \frac{1}{A} \ln \left\{ \frac{1 + \zeta_1 e^{-\beta A t}}{1 - \zeta_1 e^{-\beta A t}} \right\},$$
$$\tag{2.22}$$
$$x_2(t) = x_0 + \frac{1}{A} \ln \left\{ \frac{1 + \zeta_2 e^{-\beta A t}}{1 - \zeta_2 e^{-\beta A t}} \right\},$$

where

$$\zeta_1 = \frac{e^{-A x_0} - 1}{e^{-A x_0} + 1}, \quad \zeta_2 = \frac{e^{A(1 - x_0)} - 1}{e^{A(1 - x_0)} + 1}. \tag{2.23}$$

In the case where the following condition is satisfied:

$$1 \ll \beta A t \ll \beta A t_{\text{char}} \tag{2.24}$$

where t_{char} is the characteristic diffusion time for the buffer, we have for $x_1(t), x_2(t)$

$$x_1(t) = x_0 + \frac{2\zeta_1}{A} e^{-\beta A t} + O(e^{-2\beta A t}),$$
$$\tag{2.25}$$
$$x_2(t) = x_0 + \frac{2\zeta_2}{A} e^{-\beta A t} + O(e^{-2\beta A t}).$$

The width of the focused zone is determined by the relation

$$\delta(t) \equiv x_2(t) - x_1(t) = \frac{2}{A} e^{-\beta A t} (\zeta_2 - \zeta_1) + O(e^{-2\beta A t}). \tag{2.26}$$

We note that the segment $[x_1(t), x_2(t)]$ in this case is located asymmetrically relative to the point x_0. When $A > 0$, we have $\xi_2 > \xi_1$; when $A < 0$, $\xi_2 < \xi_1$. The approximate form of the functions $x_1(t), x_2(t), \delta(t)$ is given in Figs. 50 and 51.

3. Asymptotic Solution for Low Diffusion

Let us consider Eq. (1.9) with boundary conditions (1.11) and initial condition (1.10), assuming for simplicity $D_4 = 1$ [see (1.8)]:

$$\xi_t - \mu\xi_{xx} + f_x(x)\xi + f(x)\xi_x = 0, \tag{3.1}$$

$$i|_{x=0,1} \equiv (-\mu\xi_x + f(x)\xi)|_{x=0,1} = 0, \tag{3.2}$$

$$\xi(x, t)|_{t=0} = p(x). \tag{3.3}$$

Let the conditions (2.10) be satisfied. Then

$$f(x_0) = 0, \quad f'(x_0) < 0, \quad 0 < x_0 < 1. \tag{3.4}$$

Let us make the change of variables

$$x = x_0 + \frac{\sqrt{\mu}}{\alpha}s, \quad \alpha = \sqrt{-\frac{f'(x_0)}{2}}. \tag{3.5}$$

In the new variables, Eq. (3.1) has the form

$$\xi_t - \alpha^2\xi_{ss} + f_x\left(x_0 + \frac{\sqrt{\mu}}{\alpha}s\right)\xi + \frac{\alpha}{\sqrt{\mu}}f\left(x_0 + \frac{\sqrt{\mu}}{\alpha}s\right)\xi_s = 0. \tag{3.6}$$

Conditions (3.2), (3.3) will be the following:

$$\left\{-\alpha\sqrt{\mu}\xi_s + f\left(x_0 + \frac{\sqrt{\mu}}{\alpha}s\right)\xi\right\}_{s=\frac{\alpha(1-x_0)}{\sqrt{\mu}},\ -\frac{\alpha x_0}{\sqrt{\mu}}}, \tag{3.7}$$

$$\xi\left(x_0 + \frac{\sqrt{\mu}}{\alpha}s,\ t\right)\bigg|_{t=0} = p\left(x_0 + \frac{\sqrt{\mu}}{\alpha}s\right) \equiv p(s). \tag{3.8}$$

Let us assume that $f(x)$ can be represented as a Taylor series about the point $x = x_0$:

$$f(x) = f'(x_0)(x - x_0) + \frac{1}{2}f''(x_0)(x - x_0)^2 + \dots, \tag{3.9}$$

$$f'(x) = f'(x_0) + f''(x_0)(x - x_0) + \dots$$

Making substitution (3.5), we obtain

$$f\left(x_0 + \frac{\sqrt{\mu}}{\alpha}s\right) = f'(x_0)\frac{\sqrt{\mu}}{\alpha}s + \frac{1}{2}f''(x_0)\frac{\mu}{\alpha^2}s^2 + \dots, \tag{3.10}$$

$$f_x\left(x_0 + \frac{\sqrt{\mu}}{\alpha}s\right) = f'(x_0) + f''(x_0)\frac{\sqrt{\mu}}{\alpha}s.$$

Let

$$\mu \to 0. \tag{3.11}$$

Let us replace boundary conditions (3.7) for $\mu \to 0$ by the following:

$$\xi|_{s=\pm\infty} = 0. \tag{3.12}$$

Substituting (3.10) into (3.6) and discarding terms of order $O(\sqrt{\mu})$, $\mu \to 0$, we derive

$$\xi_t - \alpha^2 \xi_{ss} + f'(x_0)\xi + sf'(x_0)\xi_s = 0. \tag{3.13}$$

Solution (3.13) with conditions (3.12) has the form

$$\xi(s, t) = \sum_{k=0}^{\infty} e^{-\lambda_k t} a_k u_k(s). \tag{3.14}$$

Here, λ_k are the eigenvalues, and $u_k(s)$ are their corresponding eigenfunctions for the following problem:

$$-\lambda u - \alpha^2 u'' + f'(x_0) u + sf'(x_0) u' = 0, \tag{3.15}$$

$$u(\pm\infty) = 0. \tag{3.16}$$

Making one more substitution

$$u(s) = y(s) e^{-\frac{1}{2}s^2}, \tag{3.17}$$

we obtain

$$y''(s) - (1 + s^2) y(s) + \left(2 + \frac{\lambda}{\alpha^2}\right) y(s) = 0; \tag{3.18}$$

$$y(\pm\infty) = 0. \tag{3.19}$$

The eigenvalues λ_k and their corresponding eigenfunctions $\varphi_k(s)$ are well known (see [21, 26]):

$$\lambda_k = 2\alpha^2 k, \quad k = 0, 1, \ldots \ (\lambda_k = -f'(x_0) k), \tag{3.20}$$

$$\varphi_k(s) = (2^k k! \sqrt{\pi})^{-1/2} e^{-\frac{1}{2}s^2} H_k(s), \quad k = 0, 1, \ldots, \tag{3.21}$$

where $H_k(s)$ are the Chebyshev–Hermite polynomials

$$H_k(s) = (-1)^k e^{s^2} \frac{d^k}{ds^k} e^{-s^2}. \tag{3.22}$$

Now solution (3.14) takes on the form

$$\xi(s, t) = \sum_{k=0}^{\infty} e^{kf'/(x_0)t} \cdot a_k (2^k k! \sqrt{\pi})^{-1/2} e^{-s^2} H_k(s). \tag{3.23}$$

In this case, the coefficients a_k are determined by the expression

$$a_k = \int_{-\infty}^{\infty} \xi(s, 0) (2^k k! \sqrt{\pi})^{\frac{1}{2}} H_k(s) \, ds. \tag{3.24}$$

Taking into account the fact that $\mu \to 0$, this expression evidently may be replaced [using (3.5)] by the following:

$$a_k = \int_0^1 p(x) (2^k k! \sqrt{\pi})^{\frac{1}{2}} H_k \left(\frac{(x - x_0) \alpha}{\sqrt{\mu}} \right) \frac{\alpha}{\sqrt{\mu}} \, dx. \tag{3.25}$$

Taking into account (3.5), we obtain from (3.23)

$$\xi(x, t) = \sum_{k=0}^{\infty} e^{kf'(x_0)t} \sqrt{-\frac{f'(x_0)}{2}} \cdot \exp \left\{ \frac{(x - x_0)^2 f'(x_0)}{2\mu} \right\}$$

$$\times \int_0^1 p(z) H_k \left(\frac{(z - x_0) \alpha}{\sqrt{\mu}} \right) dz. \tag{3.26}$$

Finally, for the principal term of the asymptotic as $\mu \to 0$, $t \to \infty$, we have

$$\xi(x, t) \sim \xi_{st}(x) + 4e^{-f'(x_0)t} (x - x_0) \left(-\frac{f'(x_0)}{2\mu} \right)^{3/2} e^{\frac{(x-x_0)^2 f'(x_0)}{2\mu}}$$

$$\times \int_0^1 p(z) (z - x_0) \, dz, \tag{3.27}$$

$$\xi_{st}(x) = \sqrt{-\frac{f'(x_0)}{2\mu}} \exp \left\{ \frac{(x - x_0)^2 f'(x_0)}{2\mu} \right\} \cdot \int_0^1 p(z) \, dz. \tag{3.28}$$

Here, $\xi_{st}(x)$ is the steady-state solution to problem (3.1) (compare with Chapter VIII).

Analysis of expression (3.27) allows us to draw the somewhat unexpected conclusion that the quantity $\xi(x, t) - \xi_{st}(x)$ is not necessarily positive; i.e., the sample concentration in any half-neighborhood of the isoelectric point x_0 will not always be highest in the steady state.

CONCLUSION

In concluding this book, we would like to make some summaries and give our opinion concerning the prospects for further development of the theory. We might hope that the material presented may serve as a basis for developing the macroscopic theory of electromigration, making it possible to describe the motion in an electric field of a substance consisting of individual molecules, dispersed particles, cell organelles, and whole cells. This requires refinement of our information concerning the equations of state for the supporting media, in particular taking into account the plastic properties of different gels and their interaction with an electric field and with the substances to be separated.

We need to more completely investigate the possible chemical transformations for protein molecules in solution, especially the dependence of dissociation constants on concentration. We should emphasize that the theory presented in Chapter II is valid only in the case where the activity coefficients may be replaced by the rate constants.

Although at present we have neither a statistical nor a thermodynamic theory of biological phenomena, and creation of such a theory is a matter for the indefinitely far future, undoubtedly there is interest in the investigation of the interconnection between micro and macro levels of the phenomena. We refer to the development of a theoretical approach connected with consideration of individual particles of the medium. It remains to be determined as to which fraction of the information on the micro level is preserved when the description is condensed as a result of the transition to the phenomenological approach. It is possible that establishing such interconnections (even in the simplest models) will promote important progress in the theory of biological systems. Thus, from the electrophoretic separation pattern in principle we may draw conclusions concerning the change in spatial conformation of macromolecules.

Processes occurring in cells, tissues, and organisms as a whole are very much analogous to those processes considered in this book. The observed phenomena frequently may be connected with transport of substances participating in chemical reactions upon action of an electric field. Therefore, the base model presented in Chapter I, together with its subsequent concrete specification of cases, may serve as a basis for the effective use of the ideas and methods of the physics of continuous media in biology.

Most examples considered in this book refer to the one-dimensional case. It is quite obvious that there is interest in investigating two-dimensional and three-dimensional distributions of mixtures; such investigations may be done both numerically (finite-difference method) and by studying the stability of one-dimensional states and the generation of dissipative structures from them. This problem is also important in connection with so-called two-dimensional electrophoresis [71] or instruments with an electrophoretic chamber having a more complicated configuration [33, 48, 74, 77, 91].

The general equations given in this book allow us to investigate various types of convection arising upon electrophoresis. It is understood that convection leads to disruption of the one-dimensional character of the distributions and to corresponding complication of the model. For hydrodynamics, there is interest in investigating the effect of chemical reactions on gravitational and electrical convection. The results obtained in this case may find application in chemical technology. Let us also recall that there are types of electrophoresis which make use of convection [77].

There are interesting analogies between the different physicochemical methods (electrophoresis, ultracentrifugation, chromatography). Thus, in Chapter IV we pointed out a profound similarity between isotachophoresis and sorption chromatography. Obviously, a whole series of complicated cases of sedimentation and chromatographic analysis may be investigated in a way analogous to what has been demonstrated in this book for different electrophoresis models.

We hope that the new class of mathematical models describing infinite-component mixtures and allowing us to represent complicated chemical processes as motion in so-called sorting space (or S space) have prospects for further development. These models are a natural means for describing polymerization processes both in chemistry and in biology.

REFERENCES

1. A. I. Akhiezer and S. V. Peletminskii, *Methods of Statistical Physics* [in Russian], Nauka, Moscow (1977).
2. G. M. Bartenev and Yu. V. Zelenev, *Course in the Physics of Polymers* [in Russian], Khimiya, Leningrad (1976).
3. I. F. Bakhareva, *Nonlinear Nonequilibrium Thermodynamics* [in Russian], Izd. Sarat. Univ., Saratov (1976).
4. H. Bateman and A. Erdélyi, *Higher Transcendental Functions*, McGraw-Hill, New York (1953).
5. M. K. Bologa, F. P. Grosu, and I. A. Kozhukhar', *Electrical Convection and Heat Exchange* [in Russian], Shtiintsa, Kishinev (1977).
6. U. Weser, "Chemistry and structure of borate complexes with some important polyhydroxy compounds," in: *Structure and Bonding*, J. D. Dunitz et al. (eds.) [Russian translation], Mir, Moscow (1969), pp. 251–272.
7. M. M. Viktorov, *Methods of Calculating Physicochemical Quantities and Applied Calculations* [in Russian], Khimiya, Leningrad (1977).
8. S. A. Vol'fson and N. S. Enikolopyan, *Calculations of High-Efficiency Polymerization Processes* [in Russian], Khimiya, Moscow (1980).
9. V. L. Ginzburg, *Theoretical Physics and Astrophysics* [in Russian], Nauka, Moscow (1975).
10. P. Glansdorff and I. Prigogine, *Thermodynamic Theory of Structure, Stability, and Fluctuations*, Wiley, New York (1971).
11. S. R. de Groot and P. Mazur, *Nonequilibrium Thermodynamics*, North-Holland, Amsterdam (1962).
12. D. Dobos, *Electrochemical Data*, Elsevier, New York (1975).
13. S. S. Dukhin and B. V. Deryagin, *Electrophoresis* [in Russian], Nauka, Moscow (1976).

14. I. Gyarmati, *Nonequilibrium Thermodynamics*, Springer-Verlag, New York (1970).
15. M. Yu. Zhukov and V. I. Yudovich, "Basic equations of hydroelectrothermodynamics for multicomponent liquids," *Mol. Biol.*, **28**, 43–53 (1981).
16. M. Yu. Zhukov and V. I. Yudovich, "Multicomponent mixtures in local chemical equilibrium," *Mol. Biol.*, **28**, 54–57 (1981).
17. M. Yu. Zhukov and V. I. Yudovich, "Creation of a pH gradient in solution using carrier ampholytes," *Mol. Biol.*, **28**, 71–74 (1981).
18. D. N. Zubarev, *Nonequilibrium Statistical Thermodynamics* [in Russian], Nauka, Moscow (1971).
19. A. A. Il'yushin, *Mechanics of Continuous Media* [in Russian], Izd. Mosk. Univ., Moscow (1977).
20. E. A. Kaimakov and N. L. Varshavskaya, "Variation in transport numbers in aqueous solutions of electrolytes," *Usp. Khim.*, **35**, No. 2, 201–228 (1966).
21. E. Kamke, *Handbook of Ordinary Differential Equations* [Russian translation], Nauka, Moscow (1971).
22. R. L. Kay, "Measurement of transport numbers," in: *New Instrumental Measurement Methods in Electrochemistry*, Krieger, Florida (1980).
23. J. Koryta, J. Dvorak, and V. Boháčkova, *Electrochemistry*, Halsted Press, New York (1970).
24. N. S. Koshlyakov, É. B. Gliner, and M. M. Smirnov, *Partial Differential Equations of Mathematical Physics* [in Russian], Nauka, Moscow (1970).
25. N. N. Kuznetsov, "Several mathematical problems in chromatography," *Vychislit. Methody Programm.*, No. 6, 242–258 (1967).
26. R. Courant and D. Hilbert, *Methods of Mathematical Physics*, Vol. 1, Wiley, New York (1953).
27. Ya. I. Gerasimov (ed.), *Course in Physical Chemistry*, Vol. 2 [in Russian], Khimiya, Moscow (1973).
28. E. Lightfoot, *Transport Phenomena in Living Systems*, Wiley, New York (1974).
29. L. D. Landau and E. M. Lifshitz, *Statistical Physics*, Pergamon Press, New York (1959).
30. L. D. Landau and E. M. Lifshitz, *Fluid Mechanics*, Pergamon Press, New York (1959).
31. L. D. Landau and E. M. Lifshitz, *Electrodynamics of Continuous Media*, Pergamon Press, New York (1960).
32. L. D. Landau and E. M. Lifshitz, *The Classical Theory of Fields*, Pergamon Press, New York (1962).
33. É. G. Larskii, *Methods of Zone Electrophoresis* [in Russian], Meditsina, Moscow (1971).

34. V. Levich, *Physicochemical Hydrodynamics* [in Russian], Nauka, Moscow (1959).

35. *Methods of Practical Biology* [Russian translation], Mir, Moscow (1978).

36. A. P. Murel', "Synthesis and properties of carrier-ampholytes," in: *Electrophoretic Methods of Protein Analysis* [in Russian], Nauka, Novosibirsk (1981), pp. 106–114.

37. L. A. Osterman, *Investigation Methods for Proteins and Nucleic Acids. Electrophoresis and Ultracentrifugation (Practical Guide)* [in Russian], Nauka, Moscow (1981).

38. I. Prigogine, *Introduction to the Thermodynamics of Irreversible Processes*, C. C. Thomas, Springfield (1955).

39. T. M. Rice, J. C. Hansel, T. G. Phillips, and G. A. Thomas, *Electron–Hole Liquid in Semiconductors* [Russian translation], Mir, Moscow (1980).

40. B. L. Rozhdestvenskii, "Discontinuous solutions to systems of quasilinear equations of the hyperbolic type," *Usp. Mat. Nauk*, **15**, No. 6, 95–117 (1960).

41. B. L. Rozhdestvenskii and N. N. Yanenko, *Systems of Quasilinear Equations* [in Russian], Nauka, Moscow (1978).

42. L. I. Sedov, *A Course in Continuum Mechanics*, Noordhoff, Groningen (1972).

43. L. I. Sedov, *Reflections on Science and Scientists* [in Russian], Nauka, Moscow (1980).

44. J. C. Slattery, *Momentum, Energy, and Mass Transfer in Continua*, Krieger, New York (1978).

45. V. V. Stepanov, *Course in Differential Equations* [in Russian], Nauka, Moscow (1958).

46. A. V. Stepanov and E. K. Korchemnaya, *Electromigration Method in Inorganic Analysis* [in Russian], Khimiya, Moscow (1979).

47. A. Ya. Strongin, E. D. Levin, and V. M. Stepanov, "Isotachophoresis as a method for biopolymer separation," *Bioorg. Khim.*, No. 2, 869–884 (1976).

48. G. V. Troitskii, *Electrophoresis of Proteins* [in Russian], Izd. Khar'kov. Univ., Khar'kov (1962).

49. G. V. Troitskii, "Characteristic features of different isoelectric focusing methods," *Mol. Biol.*, **28**, 57–63 (1981).

50. G. V. Troitskii, G. Yu. Azhitskii, S. N. Bagdasar'yan, et al., "Study of the structure and mutability of serum albumin in the normal state and in pathology," *Mol. Biol.*, **12**, 89–98 (1976).

51. G. V. Troitskii, V. P. Zav'yalov, and V. M. Abramov, "Creation of a stable pH gradient in a buffer solution–nonelectrolyte mixture: Use of this system for isoelectric focusing of albumin and hemoglobin," *Dokl. Akad. Nauk SSSR*, **214**, No. 4, 955–958 (1974).

52. C. Truesdell, *A First Course in Rational Continuum Mechanics*, Academic Press, New York (1977).

53. M. V. Fedoryuk, *Method of Steepest Descent* [in Russian], Nauka, Moscow (1977).

54. R. Haase, *Thermodynamics of Irreversible Processes*, Addison-Wesley, Reading, Massachusetts (1968).

55. E. M. Shvarts, *Complex Compounds of Boron with Polyhydroxy Compounds* [in Russian], Zinatne, Riga (1968).

56. H. E. Avery, *Basic Reaction Kinetics and Mechanisms*, Crane Russak, New York (1974).

57. V. P. Shvedov (ed.), *Electromigration Method in Physicochemical and Radiochemical Investigations* [in Russian], Atomizdat, Moscow (1971).

58. K. B. Yatsimirskii, E. E. Kriss, and V. L. Gvyarzdovskaya, *Stability Constants for Complexes of Metals with Bioligands* [in Russian], Naukova Dumka, Kiev (1979).

59. R. J. Aitkin and R. F. Craine, "Continuum theories of mixtures: basic theory and historical development," *Q. J. Mech. Appl. Math.*, **29**, No. 2, 290–244 (1976).

60. A. Baldesten, "Theoretical and practical aspects of isotachophoretic separations of weak electrolytes," *Sci. Tools*, **27**, No. 1, 2–7 (1980).

61. J. C. Bearden, Jr., "Electrophoretic mobility of high-molecular-weight, double-stranded DNA on agarose gels," *Gene*, **6**, No. 1, 221–234 (1979).

62. R. Bishop, "Current major application areas in electrofocusing," *Sci. Tools*, **26**, No. 1, 2–8 (1979).

63. J. Bours, "Isoelectric focusing in free solution," in: *Isoelectric Focusing*, Academic Press, New York (1976), pp. 209–228.

64. R. K. Brown, M. L. Caspers, J. M. Lull, et al., "Carrier ampholyte distribution in isoelectric focusing," *J. Chromatogr.*, **131**, No. 1, 223–232 (1977).

65. J. R. Cann, D. I. Stimpson, and D. J. Cox, "Isoelectric focusing of interacting systems. III. Carrier ampholyte-induced macromolecular association or dissociation into subunits," *Anal. Biochem.*, **86**, No. 1, 34–39 (1978).

66. M. L. Caspers and A. Chrambach, "Natural pH gradients formed by amino acids: ampholyte distribution, time course, use in electrofocusing of protein. Relation to pH gradients in isotachophoresis and separator effects," *Anal. Biochem.*, **81**, No. 1, 28–39 (1977).

67. A. Chrambach and G. Baumann, "Isoelectric focusing on polyacrylamide gel," in: *Isoelectric Focusing*, Academic Press, New York (1976), pp. 77–91.

68. M. Dishon and G. H. Weiss, "When do transient double peaks oc-

cur in pH gradient electrophoresis?" *Anal. Biochem.*, **81**, No. 1, 1–9 (1977).

69. H. Eisenberg, "Sedimentation in the ultracentrifuge and diffusion of macromolecules carrying electrical charges," *Biophys. Chem.*, **5**, No. 1, 243–251 (1976).

70. R. T. Espejo and J. Lebowitz, "A simple electrophoretic method for the determination of superhelix density of closed circular DNAs and for observation of their superhelix density heterogeneity," *Anal. Biochem.*, **72**, No. 1, 95–103 (1976).

71. P. H. O'Farrell, "High resolution two-dimensional electrophoresis of proteins," *J. Biol. Chem.*, **250**, No. 10, 4007-4021 (1975).

72. M. P. Fisher and C. W. Dingman, "Role of molecular conformation in determining the electrophoretic properties of polynucleotides in agarose–acrylamide composite gels," *Biochemistry*, **10**, No. 10, 1895–1899 (1971).

73. D. H. Flint and R. E. Harrington, "Gel electrophoresis of deoxyribonucleic acid," *Biochemistry*, **11**, No. 25, 4858–4863 (1972).

74. Ö. Gaál, G. A. Medgyesi, and L. Vereczkey, *Electrophoresis in the Separation of Biological Macromolecules*, Akad. Kiadó, Budapest (1980).

75. W. J. Gelsema and C. L. de Ligny, "Isoelectric focusing as a method for the characterization of ampholytes," *J. Chromatogr.*, **130**, No. 1, 41–50 (1977).

76. H. Haglund, "Isotachophoresis," *Sci. Tools*, **17**, No. 1, 2–13 (1970).

77. H. Haglund, "Isoelectric focusing in pH gradients – a technique for fractionation and characterization of ampholytes," *Meth. Biochem. Anal.*, **19**, 1–104 (1971).

78. D. L. Hare, D. I. Stimpson, and J. R. Cann, "Multiple bands produced by interaction of a single macromolecule with carrier ampholytes during isoelectric focusing," *Arch. Biochem. Biophys.*, **187**, No. 1, 274–275 (1978).

79. E. H. Harley, J. S. White, and K. R. Rees, "The identification of different structural classes of nucleic acids by electrophoresis in polyacrylamide gels of different concentration," *Biochim. Biophys. Acta*, **299**, No. 1, 253–263 (1973).

80. P. H. Johnson and L. I. Grossman, "Electrophoresis of DNA in agarose gels: Optimizing separations of conformational isomers of double- and single-stranded DNAs," *Biochemistry*, **16**, No. 12, 4217–4225 (1977).

81. W. Keller, "Determination of the number of superhelical turns in simian virus 40 DNA by gel electrophoresis," *Proc. Natl. Acad. Sci. USA*, **72**, No. 12, 4876–4880 (1975).

82. P. D. Kelly, "Reacting continuum," *Int. J. Eng. Sci.*, **2**, No. 1, 129–153 (1964).

83. J. Linney, A. Chrambach, and D. Rodbard, "Factors affecting res-
     olution, band width, number of theoretical plates, and apparent dif-
     fusion coefficient in polyacrylamide gel electrophoresis," *Anal.
     Biochem.*, **78**, No. 1, 287–294 (1976).

84. D. Malamud and J. W. Drysdale, "Isoelectric points of proteins: a
     table," *Anal. Biochem.*, **86**, No. 2, 620–648 (1978).

85. G. T. Moore, "Theory of isotachophoresis. Development of con-
     centration boundaries," *J. Chromatogr.*, **106**, No. 1, 1–16 (1975).

86. H. P. Mühlmann and H. Schönert, "On the determination of
     association constants of proteins by electrophoretic measurements,"
     Biophys. Chem., **9**, No. 1, 149–155 (1979).

87. N. Y. Nguyen and A. Chrambach, "Stabilization of pH gradients in
     buffer electrofocusing on polyacrylamide gel," *Anal. Biochem.*,
     82, No. 1, 54–62 (1977).

88. N. Y. Nguyen and A. Chrambach, "Stabilization of pH gradients
     formed by ampholine," *Anal. Biochem.*, **82**, No. 1, 226–235
     (1977).

89. N. G. Nguyen, A. G. McCormick, and A. Chrambach, "An anodic
     drift of pH gradients in isoelectric focusing on polyacrylamide gel,"
     Anal. Biochem., **88**, No. 1, 186–196 (1978).

90. N. Y. Nguyen, A. Salokangas, and A. Chrambach, "Electrofocus-
     ing in natural pH gradients formed by buffers: gradient modifica-
     tion," *Anal. Biochem.*, **78**, No. 1, 287–294 (1977).

91. P. G. Righetti and J. W. Drysdale, *Isoelectric Focusing*, North-
     Holland, Amsterdam (1976).

92. H. Rilbe, "Isoelectric focusing – development from notion to
     practically working tool," *Sci. Tools*, **23**, No. 1, 18–21 (1976).

93. H. Rilbe, "Theoretical aspects of steady-state isoelectric focusing,"
     in: *Isoelectric Focusing*, Academic Press, New York (1976), pp.
     14–52.

94. D. Rodbard and A. Chrambach, "Estimation of molecular radius,
     free mobility, and valence using polyacrylamide gel electrophore-
     sis," *Anal. Biochem.*, **40**, No. 1, 95–134 (1971).

95. R. J. Routs, *Electrolyte Systems in Isotachophoresis and Their
     Application to Some Protein Separations*, Eindhoven (1971).

96. J. A. Schellman, "Electrical double layer, zeta potential, and elec-
     trophoretic charge of double-stranded DNA," *Biopolymers*, **16**,
     No. 8, 1415–1434 (1977).

97. R. Shinnar and G. H. Weiss, "A note on the resolution of two
     Gaussian peaks," *Separ. Sci.*, **11**, No. 4, 377–383 (1976).

98. M. Syvanen and H. K. Schachman, "Donnan effect as measured
     by sedimentation equilibrium for the protein cytochrome c," *Bio-
     polymers*, **17**, No. 4, 943–956 (1978).

99. O. Vesterberg, "The carrier ampholytes," in: *Isoelectric Focusing*,
     Academic Press, New York (1976), pp. 53–76.

100. B. S. Weir, A. H. D. Brown, and D. R. Marshall, "Testing for selective neutrality of electrophoretically detectable protein polymorphisms," *Genetics*, **84**, No. 3, 639–659 (1976).